THE ART of WAR

by

Antoine Henri de Jomini
(Baron de Jomini)
General & Aide-de-Camp of the Emperor of Russia

Originally Published in French in 1836

Tranlated from the French by

Capt. G. H. Mendell
Corps of Topographical Engineers
U.S. Army

&

Lieut. W. P. Craighill
Corps of Engineers
U.S. Army

1862

This Edition by
Arc Manor, Rockville, MD
2007

With an Introduction and Commentary by
Horace E. Cocroft, Jr.

Library of Congress Control Number: 2006936549

ISBN-10: 0-9786536-3-7

ISBN-13: 978-0-9786536-3-7

ISBN-13: 978-1-6497300-6-0

MANOR

2007

P. O. Box 10339
Rockville, MD 20849-0339
United States of America

Ordering Information

Single Copies or Bulk Orders may be ordered by contacting **Sales@ArcManor.com**. Bulk orders, as well as orders for libraries and eductional institutions are eligible for discounts.

TABLE OF CONTENTS

COMMENTARIES

Publisher's note on illustrations and text

We have retained the original illustrations to maintain the flavor of the original translation. However text has been reformatted and typeset for clarity and easy readability

PREFACE TO THE 1862 EDITION

*I*n the execution of any undertaking there are extremes on either hand which are alike to be avoided. The rule holds in a special manner in making a translation. There is, on the one side, the extreme of too rigid adherence, word for word and line for line, to the original, and on the other is the danger of using too free a pen. In either case the sense of the author may not be truly given. It is not always easy to preserve a proper mean between these extremes. The translators of Jomini's Summary of the Principles of the Art of War have endeavored to render their author into plain English, without mutilating or adding to his ideas, attempting no display and making no criticisms.

To persons accustomed to read for instruction in military matters, it is not necessary to say a word with reference to wthe merits of Jomini. To those not thus accustomed heretofore, but who are becoming more interested in such subjects, (and this class must include the great mass of the American public,) it is sufficient to say, and it may be said with entire truth, that General Jomini is admitted by all competent judges to be one of the ablest military critics and historians of this or any other day.

The translation now presented to the people has been made with the earnest hope and the sincere expectation of its proving useful. As the existence of a large, well-instructed standing army is deemed incompatible with our institutions, it becomes the more important that military information be as extensively diffused as possible among the people. If by the present work the translators shall find they have contributed, even in an inconsiderable degree, to this important object, they will be amply repaid for the care and labor expended upon it.

To those persons to whom the study of the art of war is a new one, it is recommended to begin at the article "Strategy," Chapter III., from that point to read to the end of the Second Appendix, and then to return to Chapters

I. and II. It should be borne in mind that this subject, to be appreciated, must be studied, map in hand: this remark is especially true of strategy. An acquaintance with the campaigns of Napoleon I. is quite important, as they are constantly referred to by Jomini and by all other recent writers on the military art.

U.S. Military Academy,
West Point, N.Y.
January, 1862.

INTRODUCTION
by Horace E. Cocroft, Jr.

*I*n 1991, General Norman Schwarzkopf drove Saddam Hussein out of Kuwait using several specific strategies from Jomini. Schwarzkopf established a temporary supply base in the Saudi Arabian desert to form a base of operations for the U.S. Seventh Corps and then used Marine and Arab coalition allies in a pinning operation against Iraqi troops in Kuwait while the Seventh Corps made a turning movement into the Iraqi rear. Having captured its limited, geographic objective, the coalition called a halt to the war. Schwarzkopf's strategies came straight from *The Art of War*, which is the foundation of professional military education in the Western world. Although several military writers preceded Jomini, such as Marshal Saxe in the 18th century, and some may be better known today, like Carl Von Clausewitz, no other military theoretician has had quite the impact on professional military thinking, doctrine, and vocabulary as the Swiss-born staff officer Jomini, whose work remains relevant even today.

Jomini (March 6, 1779–March 24, 1869) was born at Payerne in the canton of Vaud, Switzerland, where his father was a minor official. Jomini's early military life was spent in the armies of the French Revolution and the Helvetic Republic (the Swiss client state of the French Republic). After the 1801 Treaty of Luneville ended hostilities between Hapsburg Austria and France, Jomini relocated to Paris, ostensibly to resume the business career interrupted by war and revolution. More importantly, he wrote his first work on military theory, "Treatise on Grand Military Operations." This early work brought him to the attention of Marshal Ney, who placed Jomini on his staff, as well as the attention of Napoleon himself, who ensured Jomini's promotion to colonel and General de Brigade and awarded him the Legion of Honor. After the 1806-1807 campaigns in Prussia and Poland, Jomini rejoined Ney as his chief of staff, an arrangement that lasted until 1813.

After Napoleon's victory over the Russians and Prussians at Bautzen, Jomini quarreled with Napoleon's Chief of Staff, Louis Alexandre Berthier, and quit the French army. During the armistice that followed Bautzen, Jomini joined the Russian service, where he received the rank of lieutenant-general; based on the reputation of his writings, he won the appointment of an aide-de-camp to Czar Alexander. Jomini's postwar career was primarily in the Russian army. He was employed in the military education of Prince Nicholas and played a principle part in the organization of the Russian staff college. He retired from active service in 1829 and spent his long retirement writing and commenting on military affairs. *The Art of War* comes from the early part of his retirement.

The Art of War is a book with a purpose to educate. It is not a work of philosophy or history, although it contains useful military history and interesting comments on the philosophy of the use of armed force to achieve national goals. Its purpose focuses primarily on educating serving officers on military matters outside the regimental routines that were the bread and butter of most 19th century line officers' lives. A secondary goal of the book was to give instruction in military matters to a more general audience. Jomini succeeded in producing a book that is practical and easily comprehended. *The Art of War* is based on simple, easily understood maxims or principles. By providing commanders with a clear framework by which to plan operations, Jomini wanted to help his students eliminate the unpredictable variables involved in making war. Clausewitz, in contrast, sought to understand the fog of war rather than eliminate it.

The Art of War opens with a large perspective and then narrows it down to the details. Jomini's viewpoint progressively narrows through the general military policy of a state and grand strategies used in the fighting of a war down to the deployment of individual battalions in the line of battle. This progression—from the grand to the minute and from the general to the specific—allows him to touch upon every level of military policy in a way that manages to be both comprehensive and detailed.

The military schooling of young officers in the 19th century was an on-the-job education, usually restricted to problems at the regimental level and generally limited to leadership problems, such as keeping up troops' morale and maintaining discipline in the ranks. Battlefield tactics tended to be left to the experience and imagination of the commanding general, supplemented by any independent study in military history the general may have undertaken. What was needed was a guidebook on tactics, strategy, and logistics that provided clear, easy-to-understand principles covering most eventualities.

Jomini's *Art of War* offered a simple, practical guidebook for using troops in land operations in war. His lessons were written in such a way so as to be readily understood and copied. Considerations that were difficult to quantify, such as the morale of troops or the personalities of allied or opposing commanders, were mentioned in passing, but were generally ignored—a fact that has received the brunt of the criticism directed at Jomini. Jomini is accused of creating an instruction manual for war—just follow the checklist to ensure that the campaign will prosper. Yet an instruction manual for the strategic and operational levels of war was exactly what early 19th century Europe needed.

In the United States, Jomini is best known for the influence he had on pre-Civil War military studies at West Point. His ideas were filtered through American military interpreters like Dennis Hart Mahan and Henry Halleck. Jomini's reputation suffered at the hands of early war Union commanders as a result of the oversimplification of his thinking. Overcautious generals such as Halleck and McClellan, who used Jominian vocabulary to provide excuses for their inactivity, have been described as Jominian in their reliance on formal principles of war and cautious maneuvering for geographical advantage. The irony is that Jominian warfare was practiced by generals like Sherman, who carefully employed turning movements to maneuver Joe Johnston out of position after positioning his successful campaign to take Atlanta, and Grant, who realized that Petersburg was the nodal point that held the key to Richmond and the army of Northern Virginia. Clearly Jominian principles in the hands of imaginative generals won the war. The only major commander who could not be called Jominian was Stonewall Jackson who, in his independent operations, seemingly ignored his own lines of communication and bases of operation, relying instead on speed and surprise to create chaos in the Union army.

SUMMARY OF THE ART OF WAR

DEFINITION OF THE ART OF WAR

*T*he art of war, as generally considered, consists of five purely military branches,—viz.: Strategy, Grand Tactics, Logistics, Engineering, and Tactics. A sixth and essential branch, hitherto unrecognized, might be termed *Diplomacy in its relation to War.* Although this branch is more naturally and intimately connected with the profession of a statesman than with that of a soldier, it cannot be denied that, if it be useless to a subordinate general, it is indispensable to every general commanding an army: it enters into all the combinations which may lead to a war, and has a connection with the various operations to be undertaken in this war; and, in this view, it should have a place in a work like this.

To recapitulate, the art of war consists of six distinct parts:–

1. Statesmanship in its relation to war.

2. Strategy, or the art of properly directing masses upon the theater of war, either for defense or for invasion.

3. Grand Tactics.

4. Logistics, or the art of moving armies.

5. Engineering,—the attack and defense of fortifications.

6. Minor Tactics.

It is proposed to analyze the principal combinations of the first four branches, omitting the consideration of tactics and of the art of engineering.

Familiarity with all these parts is not essential in order to be a good infantry, cavalry, or artillery officer; but for a general, or for a staff officer, this knowledge is indispensable.

CHAPTER I

STATESMANSHIP IN ITS RELATION TO WAR

*U*nder this head are included those considerations from which a statesman concludes whether a war is proper, opportune, or indispensable, and determines the various operations necessary to attain the object of the war.

A government goes to war,—

To reclaim certain rights or to defend them;

To protect and maintain the great interests of the state, as commerce, manufactures, or agriculture;

To uphold neighboring states whose existence is necessary either for the safety of the government or the balance of power;

To fulfill the obligations of offensive and defensive alliances;

To propagate political or religious theories, to crush them out, or to defend them;

To increase the influence and power of the state by acquisitions of territory;

To defend the threatened independence of the state;

To avenge insulted honor; or,

From a mania for conquest.

It may be remarked that these different kinds of war influence in some degree the nature and extent of the efforts and operations necessary for the proposed end. The party who has provoked the war may be reduced to the defensive, and the party assailed may assume the offensive; and there may be other circumstances which will affect the nature and conduct of a war, as,—

1. A state may simply make war against another state.

2. A state may make war against several states in alliance with each other.

3. A state in alliance with another may make war upon a single enemy.

4. A state may be either the principal party or an auxiliary.

5. In the latter case a state may join in the struggle at its beginning or after it has commenced.

6. The theater of war may be upon the soil of the enemy, upon that of an ally, or upon its own.

7. If the war be one of invasion, it may be upon adjacent or distant territory: it may be prudent and cautious, or it may be bold and adventurous.

8. It may be a national war, either against ourselves or against the enemy.

9. The war may be a civil or a religious war.

War is always to be conducted according to the great principles of the art; but great discretion must be exercised in the nature of the operations to be undertaken, which should depend upon the circumstances of the case.

For example: two hundred thousand French wishing to subjugate the Spanish people, united to a man against them, would not maneuver as the same number of French in a march upon Vienna, or any other capital, to compel a peace; nor would a French army fight the guerrillas of Mina as they fought the Russians at Borodino; nor would a French army venture to march upon Vienna without considering what might be the tone and temper of the governments and communities between the Rhine and the Inn, or between the Danube and the Elbe. A regiment should always fight in nearly the same way; but commanding generals must be guided by circumstances and events.

To these different combinations, which belong more or less to statesmanship, may be added others which relate solely to the management of armies. The name Military Policy is given to them; for they belong exclusively neither to diplomacy nor to strategy, but are still of the highest importance in the plans both of a statesman and a general.

ARTICLE I

OFFENSIVE WARS TO RECLAIM RIGHTS

When a state has claims upon another, it may not always be best to enforce them by arms. The public interest must be consulted before action.

The most just war is one which is founded upon undoubted rights, and which, in addition, promises to the state advantages commensurate with the sacrifices required and the hazards incurred. Unfortunately, in our times there are so many doubtful and contested rights that most wars, though apparently based upon bequests, or wills, or marriages, are in reality but wars of expediency. The question of the succession to the Spanish crown under Louis XIV. was very clear, since it was plainly settled by a solemn will, and was supported by family ties and by the general consent of the Spanish nation; yet it was stoutly contested by all Europe, and produced a general coalition against the legitimate legatee.

Frederick II., while Austria and France were at war, brought forward an old claim, entered Silesia in force and seized this province, thus doubling the power of Prussia. This was a stroke of genius; and, even if he had failed, he could not have been much censured; for the grandeur and importance of the enterprise justified him in his attempt, as far as such attempts can be justified.

In wars of this nature no rules can be laid down. To watch and to profit by every circumstance covers all that can be said. Offensive movements should be suitable to the end to be attained. The most natural step would be to occupy the disputed territory: then offensive operations may be carried on according to circumstances and to the respective strength of the parties, the object being to secure the cession of the territory by the enemy, and the means being to threaten him in the heart of his own country. Every thing depends upon the alliances the parties may be able to secure with other states, and upon their military resources. In an offensive movement, scrupulous care must be exercised not to arouse the jealousy of any other state which might come to the aid of the enemy. It is a part of the duty of a statesman to foresee this chance, and to obviate it by making proper explanations and giving proper guarantees to other states.

ARTICLE II

OF WARS DEFENSIVE POLITICALLY, AND OFFENSIVE IN A MILITARY POINT OF VIEW

A state attacked by another which renews an old claim rarely yields it without a war: it prefers to defend its territory, as is always more honorable. But it may be advantageous to take the offensive, instead of awaiting the attack on the frontiers.

There are often advantages in a war of invasion: there are also advantages in awaiting the enemy upon one's own soil. A power with no internal dissensions, and under no apprehension of an attack by a third party, will always find it advantageous to carry the war upon hostile soil. This course will spare its territory from devastation, carry on the war at the expense of the enemy, excite the ardor of its soldiers, and depress the spirits of the adversary. Nevertheless, in a purely military sense, it is certain that an army operating in its own territory, upon a theater of which all the natural and artificial features are well known, where all movements are aided by a knowledge of the country, by the favor of the citizens, and the aid of the constituted authorities, possesses great advantages.

These plain truths have their application in all descriptions of war; but, if the principles of strategy are always the same, it is different with the political part of war, which is modified by the tone of communities, by localities, and by the characters of men at the head of states and armies. The fact of these modifications has been used to prove that war knows no rules. Military science rests upon principles which can never be safely violated in the presence of an active and skillful enemy, while the moral and political part of war presents these variations. Plans of operations are made as circumstances may demand: to execute these plans, the great principles of war must be observed.

For instance, the plan of a war against France, Austria, or Russia would differ widely from one against the brave but undisciplined bands of Turks, which cannot be kept in order, are not able to maneuver well, and possess no steadiness under misfortunes.

ARTICLE III
WARS OF EXPEDIENCY

The invasion of Silesia by Frederick II., and the war of the Spanish Succession, were wars of expediency.

There are two kinds of wars of expediency: first, where a powerful state undertakes to acquire natural boundaries for commercial and political reasons; secondly, to lessen the power of a dangerous rival or to prevent his aggrandizement. These last are wars of intervention; for a state will rarely singly attack a dangerous rival: it will endeavor to form a coalition for that purpose.

These views belong rather to statesmanship or diplomacy than to war.

ARTICLE IV
OF WARS WITH OR WITHOUT ALLIES

Of course, in a war an ally is to be desired, all other things being equal. Although a great state will more probably succeed than two weaker states in alliance against it, still the alliance is stronger than either separately. The ally not only furnishes a contingent of troops, but, in addition, annoys the enemy to a great degree by threatening portions of his frontier which otherwise would have been secure. All history teaches that no enemy is so insignificant as to be despised and neglected by any power, however formidable.

ARTICLE V
WARS OF INTERVENTION

To interfere in a contest already begun promises more advantages to a state than war under any other circumstances; and the reason is plain. The power which interferes throws upon one side of the scale its whole weight and influence; it interferes at the most opportune moment, when it can make decisive use of its resources.

There are two kinds of intervention: 1. Intervention in the internal affairs of neighboring states; 2. Intervention in external relations.

Whatever may be said as to the moral character of interventions of the first class, instances are frequent. The Romans acquired power by these interferences, and the empire of the English India Company was assured in a similar manner. These interventions are not always successful. While Russia has added to her power by interference with Poland, Austria, on the contrary, was almost ruined by her attempt to interfere in the internal affairs of France during the Revolution.

Intervention in the external relations of states is more legitimate, and perhaps more advantageous. It may be doubtful whether a nation has the right to interfere in the internal affairs of another people; but it certainly has a right to oppose it when it propagates disorder which may reach the adjoining states.

There are three reasons for intervention in exterior foreign wars,—viz.: 1, by virtue of a treaty which binds to aid; 2, to maintain the political equilibrium; 3, to avoid certain evil consequences of the war already commenced, or to secure certain advantages from the war not to be obtained otherwise.

History is filled with examples of powers which have fallen by neglect of these principles. "A state begins to decline when it permits the immoderate aggrandizement of a rival, and a secondary power may become the arbiter of nations if it throw its weight into the balance at the proper time."

In a military view, it seems plain that the sudden appearance of a new and large army as a third party in a well-contested war must be decisive. Much will depend upon its geographical position in reference to the armies already in the field. For example, in the winter of 1807 Napoleon crossed the Vistula and ventured to the walls of Koenigsberg, leaving Austria on his rear and having Russia in front. If Austria had launched an army of one hundred thousand men from Bohemia upon the Oder, it is probable that the power of Napoleon would have been ended; there is every reason to think that his army could not have regained the Rhine. Austria preferred to wait till she could raise four hundred thousand men. Two years afterward, with this force she took the field, and was beaten; while one hundred thousand men well employed at the proper time would have decided the fate of Europe.

There are several kinds of war resulting from these two different interventions:–

1. Where the intervention is merely auxiliary, and with a force specified by former treaties.

2. Where the intervention is to uphold a feeble neighbor by defending his territory, thus shifting the scene of war to other soil.

3. A state interferes as a principal party when near the theater of war,— which supposes the case of a coalition of several powers against one.

4. A state interferes either in a struggle already in progress, or interferes before the declaration of war.

When a state intervenes with only a small contingent, in obedience to treaty-stipulations, it is simply an accessory, and has but little voice in the main operations; but when it intervenes as a principal party, and with an imposing force, the case is quite different.

The military chances in these wars are varied. The Russian army in the Seven Years' War was in fact auxiliary to that of Austria and France: still, it was a principal party in the North until its occupation of Prussia. But when Generals Fermor and Soltikoff conducted the army as far as Brandenburg it acted solely in the interest of Austria: the fate of these troops, far from their base, depended upon the good or bad maneuvering of their allies.

Such distant excursions are dangerous, and generally delicate operations. The campaigns of 1799 and 1805 furnish sad illustrations of this, to which we shall again refer in Article XXIX., in discussing the military character of these expeditions.

It follows, then, that the safety of the army may be endangered by these distant interventions. The counterbalancing advantage is that its own territory cannot then be easily invaded, since the scene of hostilities is so distant; so that what may be a misfortune for the general may be, in a measure, an advantage to the state.

In wars of this character the essentials are to secure a general who is both a statesman and a soldier; to have clear stipulations with the allies as to the part to be taken by each in the principal operations; finally, to agree upon an objective point which shall be in harmony with the common interests. By the neglect of these precautions, the greater number of coalitions have failed, or have maintained a difficult struggle with a power more united but weaker than the allies.

The third kind of intervention, which consists in interfering with the whole force of the state and near to its frontiers, is more promising than the others. Austria had an opportunity of this character in 1807, but failed to profit by it: she again had the opportunity in 1813. Napoleon had just collected his forces in Saxony, when Austria, taking his front of operations in reverse, threw herself into the struggle with two hundred thousand men, with almost perfect certainty of success. She regained in two months the Italian empire and her influence in Germany, which had been lost by fifteen years of disaster. In this intervention Austria had not only the political but also the military chances in her favor,—a double result, combining the highest advantages.

Her success was rendered more certain by the fact that while the theater was sufficiently near her frontiers to permit the greatest possible display of force, she at the same time interfered in a contest already in progress, upon which she entered with the whole of her resources and at the time most opportune for her.

This double advantage is so decisive that it permits not only powerful monarchies, but even small states, to exercise a controlling influence when they know how to profit by it.

Two examples may establish this. In 1552, the Elector Maurice of Saxony boldly declared war against Charles V., who was master of Spain, Italy, and the German empire, and had been victorious over Francis I. and held France in his grasp. This movement carried the war into the Tyrol, and arrested the great conqueror in his career.

In 1706, the Duke of Savoy, Victor Amadeus, by declaring himself hostile to Louis XIV., changed the state of affairs in Italy, and caused the recall of the French army from the banks of the Adige to the walls of Turin, where it encountered the great catastrophe which immortalized Prince Eugene.

Enough has been said to illustrate the importance and effect of these opportune interventions: more illustrations might be given, but they could not add to the conviction of the reader.

ARTICLE VI

AGGRESSIVE WARS FOR CONQUEST AND OTHER REASONS

There are two very different kinds of invasion: one attacks an adjoining state; the other attacks a distant point, over intervening territory of great extent whose inhabitants may be neutral, doubtful, or hostile.

Wars of conquest, unhappily, are often prosperous,—as Alexander, Caesar, and Napoleon during a portion of his career, have fully proved. However, there are natural limits in these wars, which cannot be passed without incurring great disaster. Cambyses in Nubia, Darius in Scythia, Crassus and the Emperor Julian among the Parthians, and Napoleon in Russia, furnish bloody proofs of these truths.—The love of conquest, however, was not the only motive with Napoleon: his personal position, and his contest with England, urged him to enterprises the aim of which was to make him supreme. It is true that he loved war and its chances; but he was also a victim to the necessity of succeeding in his efforts or of yielding to England. It might be said that he was sent into this world to teach generals and statesmen what they should avoid. His victories teach what may be accomplished by activity, boldness, and skill; his disasters, what might have been avoided by prudence.

A war of invasion without good reason—like that of Genghis Khan—is a crime against humanity; but it may be excused, if not approved, when induced by great interests or when conducted with good motives.

The invasions of Spain of 1808 and of 1823 differed equally in object and in results: the first was a cunning and wanton attack, which threatened the existence of the Spanish nation, and was fatal to its author; the second, while combating dangerous principles, fostered the general interests of the country, and was the more readily brought to a successful termination because its object met with the approval of the majority of the people whose territory was invaded.

These illustrations show that invasions are not necessarily all of the same character. The first contributed largely to the fall of Napoleon; the second restored the relation between France and Spain, which ought never to have been changed.

Let us hope that invasions may be rare. Still, it is better to attack than to be invaded; and let us remember that the surest way to check the spirit of conquest and usurpation is to oppose it by intervention at the proper time.

An invasion, to be successful, must, be proportioned in magnitude to the end to be attained and to the obstacles to be overcome.

An invasion against an exasperated people, ready for all sacrifices and likely to be aided by a powerful neighbor, is a dangerous enterprise, as was well proved by the war in Spain, (1808,) and by the wars of the Revolution in 1792, 1793, and 1794. In these latter wars, if France was better prepared than Spain, she had no powerful ally, and she was attacked by all Europe upon both land and sea.

Although the circumstances were different, the Russian invasion of Turkey developed, in some respects, the same symptoms of national resistance. The religious hatred of the Ottoman powerfully incited him to arms; but the same motive was powerless among the Greeks, who were twice as numerous as the Turks. Had the interests of the Greeks and Turks been harmonized, as were those of Alsace with France, the united people would have been stronger, but they would have lacked the element of religious fanaticism. The war of 1828 proved that Turkey was formidable only upon the frontiers, where her bravest troops were found, while in the interior all was weakness.

When an invasion of a neighboring territory has nothing to fear from the inhabitants, the principles of strategy shape its course. The popular feeling rendered the invasions of Italy, Austria, and Prussia so prompt. (These military points are treated of in Article XXIX.) But when the invasion is distant and extensive territories intervene, its success will depend more upon diplomacy than upon strategy. The first step to insure success will be to secure the sincere and devoted alliance of a state adjoining the enemy, which will afford reinforcements of troops, and, what is still more important, give a secure base of operations, depots of supplies, and a safe refuge in case of disaster. The ally must have the same interest in success as the invaders, to render all this possible.

Diplomacy, while almost decisive in distant expeditions, is not powerless in adjacent invasions; for here a hostile intervention may arrest the most brilliant successes. The invasions of Austria in 1805 and 1809 might have ended differently if Prussia had interfered. The invasion of the North of Germany in 1807 was, so to speak, permitted by Austria. That of Rumelia in 1829 might have ended in disaster, had not a wise statesmanship by negotiation obviated all chance of intervention.

ARTICLE VII
WARS OF OPINION

Although wars of opinion, national wars, and civil wars are sometimes confounded, they differ enough to require separate notice.

Wars of opinion may be intestine, both intestine and foreign, and, lastly, (which, however, is rare,) they may be foreign or exterior without being intestine or civil.

Wars of opinion between two states belong also to the class of wars of intervention; for they result either from doctrines which one party desires to propagate among its neighbors, or from dogmas which it desires to crush,— in both cases leading to intervention. Although originating in religious or political dogmas, these wars are most deplorable; for, like national wars, they enlist the worst passions, and become vindictive, cruel, and terrible.

The wars of Islamism, the Crusades, the Thirty Years' War, the wars of the League, present nearly the same characteristics. Often religion is the pretext to obtain political power, and the war is not really one of dogmas. The successors of Mohammed cared more to extend their empire than to preach the Koran, and Philip II., bigot as he was, did not sustain the League in France for the purpose of advancing the Roman Church. We agree with M. Ancelot that Louis IX., when he went on a crusade in Egypt, thought more of the commerce of the Indies than of gaining possession of the Holy Sepulcher.

The dogma sometimes is not only a pretext, but is a powerful ally; for it excites the ardor of the people, and also creates a party. For instance, the Swedes in the Thirty Years' War, and Philip II. in France, had allies in the country more powerful than their armies. It may, however, happen, as in the Crusades and the wars of Islamism, that the dogma for which the war is waged, instead of friends, finds only bitter enemies in the country invaded; and then the contest becomes fearful.

The chances of support and resistance in wars of political opinions are about equal. It may be recollected how in 1792 associations of fanatics thought it possible to propagate throughout Europe the famous declaration of the rights of man, and how governments became justly alarmed, and rushed to arms probably with the intention of only forcing the lava of this volcano back into its crater and there extinguishing it. The means were not fortunate; for war and aggression are inappropriate measures for arresting an evil which lies wholly in the human passions, excited in a temporary paroxysm, of less duration as it is the more violent. Time is the true remedy for all bad passions and for all anarchical doctrines. A civilized nation may bear the yoke of a factious and unrestrained multitude for a short interval; but these storms soon pass away, and reason resumes her sway. To attempt to restrain such a mob by a foreign force is to attempt to restrain the explosion of a mine when the powder has already been ignited: it is far better to await the explosion and afterward fill up the crater than to try to prevent it and to perish in the attempt.

After a profound study of the Revolution, I am convinced that, if the Girondists and National Assembly had not been threatened by foreign armaments, they would never have dared to lay their sacrilegious hands upon the feeble but venerable head of Louis XVI. The Girondists would never have been crushed by the Mountain but for the reverses of Dumouriez and the threats of invasion. And if they had been permitted to clash and quarrel with each other to their hearts' content, it is probable that, instead of giving place to the terrible Convention, the Assembly would slowly have returned to the restoration of good, temperate, monarchical doctrines, in accordance with the necessities and the immemorial traditions of the French.

In a military view these wars are fearful, since the invading force not only is met by the armies of the enemy, but is exposed to the attacks of an exasperated people. It may be said that the violence of one party will necessarily create support for the invaders by the formation of another and opposite one; but, if the exasperated party possesses all the public resources, the armies, the forts, the arsenals, and if it is supported by a large majority of the people, of what avail will be the support of the faction which possesses no such means? What service did one hundred thousand Vendeans and one hundred thousand Federalists do for the Coalition in 1793?

History contains but a single example of a struggle like that of the Revolution; and it appears to clearly demonstrate the danger of attacking an intensely-excited nation. However the bad management of the military operations was one cause of the unexpected result, and before deducing any certain maxims from this war, we should ascertain what would have been the result if after the flight of Dumouriez, instead of destroying and capturing fortresses, the allies had informed the commanders of those fortresses that they contemplated no wrong to France, to her forts or her brave armies, and had marched on Paris with two hundred thousand men. They might have restored the monarchy; and, again, they might never have returned, at least without the protection of an equal force on their retreat to the Rhine. It is difficult to decide this, since the experiment was never made, and as all would have depended upon the course of the French nation and the army. The problem thus presents two equally grave solutions. The campaign of 1793 gave one; whether the other might have been obtained, it is difficult to say. Experiment alone could have determined it.

The military precepts for such wars are nearly the same as for national wars, differing, however, in a vital point. In national wars the country should be occupied and subjugated, the fortified places besieged and reduced, and the armies destroyed; whereas in wars of opinion it is of less importance to subjugate the country; here great efforts should be made to gain the end speedily, without delaying for details, care being constantly taken to avoid any acts which might alarm the nation for its independence or the integrity of its territory.

The war in Spain in 1823 is an example which may be cited in favor of this course in opposition to that of the Revolution. It is true that the conditions were slightly different; for the French army of 1792 was made up of more solid elements than that of the Radicals of the Isla de Leon. The war of the Revolution was at once a war of opinion, a national war, and a civil war,— while, if the first war in Spain in 1808 was thoroughly a national war, that of 1823 was a partial struggle of opinions without the element of nationality; and hence the enormous difference in the results.

Moreover, the expedition of the Duke of Angouleme was well carried out. Instead of attacking fortresses, he acted in conformity to the above-mentioned precepts. Pushing on rapidly to the Ebro, he there divided his forces, to seize, at their sources, all the elements of strength of their enemies,—which they could safely do, since they were sustained by a majority of the inhabitants. If he had followed the instructions of the Ministry, to proceed methodically to the conquest of the country and the reduction of the fortresses between the Pyrenees and the Ebro, in order to provide a base of operations, he would perhaps have failed in his mission, or at least made the war a long and bloody one, by exciting the national spirit by an occupation of the country similar to that of 1807.

Emboldened by the hearty welcome of the people, he comprehended that it was a political operation rather than a military one, and that it behooved him to consummate it rapidly. His conduct, so different from that of the allies in 1793, deserves careful attention from all charged with similar missions. In three months the army was under the walls of Cadiz.

If the events now transpiring in the Peninsula prove that statesmanship was not able to profit by success in order to found a suitable and solid order of things, the fault was neither in the army nor in its commanders, but in the Spanish government, which, yielding to the counsel of violent reactionaries, was unable to rise to the height of its mission. The arbiter between two great hostile interests, Ferdinand blindly threw himself into the arms of the party which professed a deep veneration for the throne, but which intended to use the royal authority for the furtherance of its own ends, regardless of consequences. The nation remained divided in two hostile camps, which it would not have been impossible to calm and reconcile in time. These camps came anew into collision, as I predicted in Verona in 1823,—a striking lesson, by which no one is disposed to profit in that beautiful and unhappy land, although history is not wanting in examples to prove that violent reactions, any more than revolutions, are not elements with which to construct and consolidate. May God grant that from this frightful conflict may emerge a strong and respected monarchy, equally separated from all factions, and based upon a disciplined army as well as upon the general interests of the country,—a monarchy capable of rallying to its support this incomprehensible Spanish nation, which, with merits not less extraordinary than its faults, was always a problem for those who were in the best position to know it.

ARTICLE VIII
NATIONAL WARS

National wars, to which we have referred in speaking of those of invasion, are the most formidable of all. This name can only be applied to such as are waged against a united people, or a great majority of them, filled with a noble ardor and determined to sustain their independence: then every step is disputed, the army holds only its camp-ground, its supplies can only be obtained at the point of the sword, and its convoys are everywhere threatened or captured.

The spectacle of a spontaneous uprising of a nation is rarely seen; and, though there be in it something grand and noble which commands our admiration, the consequences are so terrible that, for the sake of humanity, we ought to hope never to see it. This uprising must not be confounded with a national defense in accordance with the institutions of the state and directed by the government.

This uprising may be produced by the most opposite causes. The serfs may rise in a body at the call of the government, and their masters, affected by a noble love of their sovereign and country, may set them the example and take the command of them; and, similarly, a fanatical people may arm under the appeal of its priests; or a people enthusiastic in its political opinions, or animated by a sacred love of its institutions, may rush to meet the enemy in defense of all it holds most dear.

The control of the sea is of much importance in the results of a national invasion. If the people possess a long stretch of coast, and are masters of the sea or in alliance with a power which controls it, their power of resistance is quintupled, not only on account of the facility of feeding the insurrection and of alarming the enemy on all the points he may occupy, but still more by the difficulties which will be thrown in the way of his procuring supplies by the sea.

The nature of the country may be such as to contribute to the facility of a national defense. In mountainous countries the people are always most formidable; next to these are countries covered with extensive forests.

The resistance of the Swiss to Austria and to the Duke of Burgundy, that of the Catalans in 1712 and in 1809, the difficulties encountered by the Russians in the subjugation of the tribes of the Caucasus, and, finally, the reiterated efforts of the Tyrolese, clearly demonstrate that the inhabitants of mountainous regions have always resisted for a longer time than those of the plains,—which is due as much to the difference in character and customs as to the difference in the natural features of the countries.

Defiles and large forests, as well as rocky regions, favor this kind of defense; and the Bocage of La Vendee, so justly celebrated, proves that any country, even if it be only traversed by large hedges and ditches or canals, admits of a formidable defense.

The difficulties in the path of an army in wars of opinions, as well as in national wars, are very great, and render the mission of the general conducting them very difficult. The events just mentioned, the contest of the Netherlands with Philip II. and that of the Americans with the English, furnish evident proofs of this; but the much more extraordinary struggle of La Vendee with the victorious Republic, those of Spain, Portugal, and the Tyrol against Napoleon, and, finally, those of the Morea against the Turks, and of Navarre against the armies of Queen Christina, are still more striking illustrations.

The difficulties are particularly great when the people are supported by a considerable nucleus of disciplined troops. The invader has only an army: his adversaries have an army, and a people wholly or almost wholly in arms, and making means of resistance out of every thing, each individual of whom conspires against the common enemy; even the non-combatants have an interest in his ruin and accelerate it by every means in their power. He holds scarcely any ground but that upon which he encamps; outside the limits of his camp every thing is hostile and multiplies a thousandfold the difficulties he meets at every step.

These obstacles become almost insurmountable when the country is difficult. Each armed inhabitant knows the smallest paths and their connections; he finds everywhere a relative or friend who aids him; the commanders also know the country, and, learning immediately the slightest movement on the part of the invader, can adopt the best measures to defeat his projects; while the latter, without information of their movements, and not in a condition to send out detachments to gain it, having no resource but in his bayonets, and certain safety only in the concentration of his columns, is like a blind man: his combinations are failures; and when, after the most carefully-concerted movements and the most rapid and fatiguing marches, he thinks he is about to accomplish his aim and deal a terrible blow, he finds no signs of the enemy but his camp-fires: so that while, like Don Quixote, he is attacking windmills, his adversary is on his line of communications, destroys the detachments left to guard it, surprises his convoys, his depots, and carries on a war so disastrous for the invader that he must inevitably yield after a time.

In Spain I was a witness of two terrible examples of this kind. When Ney's corps replaced Soult's at Corunna, I had camped the companies of the artillery-train between Betanzos and Corunna, in the midst of four brigades distant from the camp from two to three leagues, and no Spanish forces had been seen within fifty miles; Soult still occupied Santiago de Compostela, the division Maurice-Mathieu was at Ferrol and Lugo, Marchand's at Corunna and Betanzos: nevertheless, one fine night the companies of the train—men and horses—disappeared, and we were never able to discover what became of them: a solitary wounded corporal escaped to report that the peasants, led by their monks and priests, had thus made away with them. Four months afterward, Ney with a single division marched to conquer the Asturias, descending the valley of the Navia, while Kellermann debouched from Leon by the Oviedo road. A part of the corps of La Romana which was guarding the Asturias marched behind the very heights which inclose the valley of the Navia, at most but a league from our columns, without the marshal knowing a word of it: when he was entering Gijon, the army of La Romana attacked the center of the regiments of the division Marchand, which, being scattered to guard Galicia, barely escaped, and that only by the prompt return of the marshal to Lugo. This war presented a thousand incidents as striking as this. All the gold of Mexico could not have procured reliable information for the French; what was given was but a lure to make them fall more readily into snares.

No army, however disciplined, can contend successfully against such a system applied to a great nation, unless it be strong enough to hold all the essential points of the country, cover its communications, and at the same time furnish an active force sufficient to beat the enemy wherever he may present himself. If this enemy has a regular army of respectable size to be a nucleus around which to rally the people, what force will be sufficient to be superior everywhere, and to assure the safety of the long lines of communication against numerous bodies?

The Peninsular War should be carefully studied, to learn all the obstacles which a general and his brave troops may encounter in the occupation or conquest of a country whose people are all in arms. What efforts of patience, courage, and resignation did it not cost the troops of Napoleon, Massena, Soult, Ney, and Suchet to sustain themselves for six years against three or four hundred thousand armed Spaniards and Portuguese supported by the regular armies of Wellington, Beresford, Blake, La Romana, Cuesta, Castanos, Reding, and Ballasteros!

If success be possible in such a war, the following general course will be most likely to insure it,—viz.: make a display of a mass of troops proportioned to the obstacles and resistance likely to be encountered, calm the popular passions in every possible way, exhaust them by time and patience, display courtesy, gentleness, and severity united, and, particularly, deal justly. The examples of Henry IV. in the wars of the League, of Marshal Berwick in Catalonia, of Suchet in Aragon and Valencia, of Hoche in La Vendee, are models of their kind, which may be employed according to circumstances with equal success. The admirable order and discipline of the armies of Diebitsch and Paskevitch in the late war were also models, and were not a little conducive to the success of their enterprises.

The immense obstacles encountered by an invading force in these wars have led some speculative persons to hope that there should never be any other kind, since then wars would become more rare, and, conquest being also more difficult, would be less a temptation to ambitious leaders. This reasoning is rather plausible than solid; for, to admit all its consequences, it would be necessary always to be able to induce the people to take up arms, and it would also be necessary for us to be convinced that there would be in the future no wars but those of conquest, and that all legitimate though secondary wars, which are only to maintain the political equilibrium or defend the public interests, should never occur again: otherwise, how could it be known when and how to excite the people to a national war? For example, if one hundred thousand Germans crossed the Rhine and entered France, originally with the intention of preventing the conquest of Belgium by France, and without any other ambitious project, would it be a case where the whole population—men, women, and children—of Alsace, Lorraine, Champagne, and Burgundy, should rush to arms? to make a Saragossa of every walled town, to bring about, by way of reprisals, murder, pillage, and incendiarism throughout the country? If all this be not done, and the Germans, in consequence of some success, should occupy these provinces, who can say that they might not afterward seek to appropriate a part of them, even though at first they had never contemplated it? The difficulty of answering these two questions would seem to argue in favor of national wars. But is there no means of repelling such an invasion without bringing about an uprising of the whole population and a war of extermination? Is there no mean between these contests between the people and the old regular method of war between permanent armies? Will it not be sufficient, for the efficient defense of the country, to organize a militia, or landwehr, which, uniformed and called by their governments into service, would regulate the part the people should take in the war, and place just limits to its barbarities?

I answer in the affirmative; and, applying this mixed system to the cases stated above, I will guarantee that fifty thousand regular French troops, supported by the National Guards of the East, would get the better of this German army which had crossed the Vosges; for, reduced to fifty thousand men by many detachments, upon nearing the Meuse or arriving in Argonne it would have one hundred thousand men on its hands. To attain this mean, we have laid it down as a necessity that good national reserves be prepared for the army; which will be less expensive in peace and will insure the defense of the country in war. This system was used by France in 1792, imitated by Austria in 1809, and by the whole of Germany in 1813.

I sum up this discussion by asserting that, without being a utopian philanthropist, or a condottieri, a person may desire that wars of extermination may be banished from the code of nations, and that the defenses of nations by disciplined militia, with the aid of good political alliances, may be sufficient to insure their independence.

As a soldier, preferring loyal and chivalrous warfare to organized assassination, if it be necessary to make a choice, I acknowledge that my prejudices are in favor of the good old times when the French and English Guards courteously invited each other to fire first,—as at Fontenoy,—preferring them to the frightful epoch when priests, women, and children throughout Spain plotted the murder of isolated soldiers.

ARTICLE IX

CIVIL WARS, AND WARS OF RELIGION

Intestine wars, when not connected with a foreign quarrel, are generally the result of a conflict of opinions, of political or religious sectarianism. In the Middle Ages they were more frequently the collisions of feudal parties. Religious wars are above all the most deplorable.

We can understand how a government may find it necessary to use force against its own subjects in order to crush out factions which would weaken the authority of the throne and the national strength; but that it should murder its citizens to compel them to say their prayers in French or Latin, or to recognize the supremacy of a foreign pontiff, is difficult of conception. Never was a king more to be pitied than Louis XIV., who persecuted a million of industrious Protestants, who had put upon the throne his own Protestant ancestor. Wars of fanaticism are horrible when mingled with exterior wars, and they are also frightful when they are family quarrels. The history of France in the times of the League should be an eternal lesson for nations and kings. It is difficult to believe that a people so noble and chivalrous in the time of Francis I. should in twenty years have fallen into so deplorable a state of brutality.

To give maxims in such wars would be absurd. There is one rule upon which all thoughtful men will be agreed: that is, to unite the two parties or sects to drive the foreigners from the soil, and afterward to reconcile by treaty the conflicting claims or rights. Indeed, the intervention of a third power in a religious dispute can only be with ambitious views.

Governments may in good faith intervene to prevent the spreading of a political disease whose principles threaten social order; and, although these fears are generally exaggerated and are often mere pretexts, it is possible that a state may believe its own institutions menaced. But in religious disputes this is never the case; and Philip II. could have had no other object in interfering in the affairs of the League than to subject France to his influence, or to dismember it.

ARTICLE X

DOUBLE WARS, AND THE DANGER OF UNDERTAKING TWO WARS AT ONCE

The celebrated maxim of the Romans, not to undertake two great wars at the same time, is so well known and so well appreciated as to spare the necessity of demonstrating its wisdom.

A government maybe compelled to maintain a war against two neighboring states; but it will be extremely unfortunate if it does not find an ally to come to its aid, with a view to its own safety and the maintenance of the political equilibrium. It will seldom be the case that the nations allied against it will have the same interest in the war and will enter into it with all their resources; and, if one is only an auxiliary, it will be an ordinary war.

Louis XIV., Frederick the Great, the Emperor Alexander, and Napoleon, sustained gigantic struggles against united Europe. When such contests arise from voluntary aggressions, they are proof of a capital error on the part of the state which invites them; but if they arise from imperious and inevitable circumstances they must be met by seeking alliances, or by opposing such means of resistance as shall establish something like equality between the strength of the parties.

The great coalition against Louis XIV., nominally arising from his designs on Spain, had its real origin in previous aggressions which had alarmed his neighbors. To the combined forces of Europe he could only oppose the faithful alliance of the Elector of Bavaria, and the more equivocal one of the Duke of Savoy, who, indeed, was not slow in adding to the number of his enemies. Frederick, with only the aid of the subsidies of England, and fifty thousand auxiliaries from six different states, sustained a war against the three most powerful monarchies of Europe: the division and folly of his opponents were his best friends.

Both these wars, as well as that sustained by Alexander in 1812, it was almost impossible to avoid.

France had the whole of Europe on its hands in 1793, in consequence of the extravagant provocations of the Jacobins, and the Utopian ideas of the Girondists, who boasted that with the support of the English fleets they would defy all the kings in the world. The result of these absurd calculations was a frightful upheaval of Europe, from which France miraculously escaped.

Napoleon is, to a certain degree, the only modern sovereign who has voluntarily at the same time undertaken two, and even three, formidable wars,—with Spain, with England, and with Russia; but in the last case he expected the aid of Austria and Prussia, to say nothing of that of Turkey and Sweden, upon which he counted with too much certainty; so that the enterprise was not so adventurous on his part as has been generally supposed.

It will be observed that there is a great distinction between a war made against a single state which is aided by a third acting as an auxiliary, and two wars conducted at the same time against two powerful nations in opposite quarters, who employ all their forces and resources. For instance, the double contest of Napoleon in 1809 against Austria and Spain aided by England was a very different affair from a contest with Austria assisted by an auxiliary force of a given strength. These latter contests belong to ordinary wars.

It follows, then, in general, that double wars should be avoided if possible, and, if cause of war be given by two states, it is more prudent to dissimulate or neglect the wrongs suffered from one of them, until a proper opportunity for redressing them shall arrive. The rule, however, is not without exception: the respective forces, the localities, the possibility of finding allies to restore, in a measure, equality of strength between the parties, are circumstances which will influence a government so threatened. We now have fulfilled our task, in noting both the danger and the means of remedying it.

COMMENTARY ON CHAPTER I

*C*hapter 1 defines wars and the relationship of diplomacy and statesmanship in the conduct of military operations, probably owing a great deal to Clausewitz. There is no hint of this breadth of thought in the "Treatise on Grand Military Operations" of Jomini's early career. His concession that different types of wars need to be fought in differing fashions is an echo of Clausewitz's famous writings in On War about the impact of the political sphere on the military. The categorization of different types of wars is somewhat arbitrary, but the advice in each article is of importance to both soldier and statesman. In this chapter, Jomini issues warnings about paying proper attention to the general diplomatic situation and to international opinion; public opinion is not ignored. Knowledge of domestic opinion is important so that leaders have an understanding of the level of sacrifice their people will be willing to accept. According to Jomini, an accurate assessment of popular opinion in enemy states is vital in evaluating the level of opposition that can be expected during war. These warnings provide useful reminders that wars aren't fought in a vacuum. Jomini gives concrete examples of how differing political situations affect military operations. In Article VI, he contrasts Napoleon's futile attempt to conquer Spain in 1808 with the successful French intervention in Spain's internal affairs in 1823 to show how the different internal political climate of Spain required different military strategies and produced different outcomes. Politicians and statesmen cannot simply hand over the conduct of a war to generals without apprising them of the general diplomatic and political situation and cautioning them as to any restraints on military operations that the situation demands.

CHAPTER II

MILITARY POLICY

*W*e have already explained what we understand by this title. It embraces the moral combinations relating to the operations of armies. If the political considerations which we have just discussed be also moral, there are others which influence, in a certain degree, the conduct of a war, which belong neither to diplomacy, strategy, nor tactics. We include these under the head of *Military Policy*.

Military policy may be said to embrace all the combinations of any projected war, except those relating to the diplomatic art and strategy; and, as their number is considerable, a separate article cannot be assigned to each without enlarging too much the limits of this work, and without deviating from my intention,—which is, not to give a treatise on theses subjects, but to point out their relations to military operations.

Indeed, in this class we may place the passions of the nation to be fought, their military system, their immediate means and their reserves, their financial resources, the attachment they bear to their government or their institutions, the character of the executive, the characters and military abilities of the commanders of their armies, the influence of cabinet councils or councils of war at the capital upon their operations, the system of war in favor with their staff, the established force of the state and its armament, the military geography and statistics of the state which is to be invaded, and, finally, the resources and obstacles of every kind likely to be met with, all of which are included neither in diplomacy nor in strategy.

There are no fixed rules on such subjects, except that the government should neglect nothing in obtaining a knowledge of these details, and that it is indispensable to take them into consideration in the arrangement of all plans. We propose to sketch the principal points which ought to guide in this sort of combinations.

ARTICLE XI
MILITARY STATISTICS AND GEOGRAPHY

By the first of these sciences we understand the most thorough knowledge possible of the elements of power and military resources of the enemy with whom we are called upon to contend; the second consists in the topographical and strategic description of the theater of war, with all the obstacles, natural or artificial, to be encountered, and the examination of the permanent decisive points which may be presented in the whole extent of the frontier or throughout the extent of the country. Besides the minister of war, the commanding general and his chief of staff should be afforded this information, under the penalty of cruel miscalculations in their plans, as happens frequently in our day, despite the great strides civilized nations have taken in statistical, diplomatic, geographical, and topographical sciences. I will cite two examples of which I was cognizant. In 1796, Moreau's army, entering the Black Forest, expected to find terrible mountains, frightful defiles and forests, and was greatly surprised to discover, after climbing the declivities of the plateau that slope to the Rhine, that these, with their spurs, were the only mountains, and that the country, from the sources of the Danube to Donauwerth, was a rich and level plain.

The second example was in 1813. Napoleon and his whole army supposed the interior of Bohemia to be very mountainous,—whereas there is no district in Europe more level, after the girdle of mountains surrounding it has been crossed, which may be done in a single march.

All European officers held the same erroneous opinions in reference to the Balkan and the Turkish force in the interior. It seemed that it was given out at Constantinople that this province was an almost impregnable barrier and the palladium of the empire,—an error which I, having lived in the Alps, did not entertain. Other prejudices, not less deeply rooted, have led to the belief that a people all the individuals of which are constantly armed would constitute a formidable militia and would defend themselves to the last extremity. Experience has proved that the old regulations which placed the elite of the Janissaries in the frontier-cities of the Danube made the population of those cities more warlike than the inhabitants of the interior. In fact, the projects of reform of the Sultan Mahmoud required the overthrow of the old system, and there was no time to replace it by the new: so that the empire was defenseless. Experience has constantly proved that a mere multitude of brave men armed to the teeth make neither a good army nor a national defense.

Let us return to the necessity of knowing well the military geography and statistics of an empire. These sciences are not set forth in treatises, and are yet to be developed. Lloyd, who wrote an essay upon them, in describing the frontiers of the great states of Europe, was not fortunate in his maxims and predictions. He saw obstacles everywhere; he represents as impregnable the Austrian frontier on the Inn, between the Tyrol and Passau, where Napoleon and Moreau maneuvered and triumphed with armies of one hundred and fifty thousand men in 1800, 1805, and 1809.

But, if these sciences are not publicly taught, the archives of the European staff must necessarily possess many documents valuable for instruction in them,—at least for the special staff school. Awaiting the time when some studious officer, profiting by those published and unpublished documents, shall present Europe with a good military and strategic geography, we may, thanks to the immense progress of topography of late years, partially supply the want of it by the excellent charts published in all European countries within the last twenty years. At the beginning of the French Revolution topography was in its infancy: excepting the semi-topographical map of Cassini, the works of Bakenberg alone merited the name. The Austrian and Prussian staff schools, however, were good, and have since borne fruit. The charts published recently at Vienna, at Berlin, Munich, Stuttgart, and Paris, as well as those of the institute of Herder at Fribourg, promise to future generals immense resources unknown to their predecessors.

Military statistics is not much better known than geography. We have but vague and superficial statements, from which the strength of armies and navies is conjectured, and also the revenue supposed to be possessed by a state,—which is far from being the knowledge necessary to plan operations. Our object here is not to discuss thoroughly these important subjects, but to indicate them, as facilitating success in military enterprises.

ARTICLE XII

OTHER CAUSES WHICH EXERCISE AN INFLUENCE UPON THE SUCCESS OF A WAR

As the excited passions of a people are of themselves always a powerful enemy, both the general and his government should use their best efforts to allay them. We have nothing to add to what has been said on this point under the head of national wars.

On the other hand, the general should do every thing to electrify his own soldiers, and to impart to them the same enthusiasm which he endeavors to repress in his adversaries. All armies are alike susceptible of this spirit: the springs of action and means, only, vary with the national character. Military eloquence is one means, and has been the subject of many a treatise. The proclamations of Napoleon and of Paskevitch, the addresses of the ancients to their soldiers, and those of Suwaroff to men of still greater simplicity, are models of their different kinds. The eloquence of the Spanish Juntas, and the miracles of the Madonna del Pilar, led to the same results by very different means. In general, a cherished cause, and a general who inspires confidence by previous success, are powerful means of electrifying an army and conducing to victory. Some dispute the advantages of this enthusiasm, and prefer imperturbable coolness in battle. Both have unmistakable advantages and disadvantages. Enthusiasm impels to the performance of great actions: the difficulty is in maintaining it constantly; and, when discouragement succeeds it, disorder easily results.

The greater or less activity and boldness of the commanders of the armies are elements of success or failure, which cannot be submitted to rules. A cabinet and a commander ought to consider the intrinsic value of their troops, and that resulting from their organization as compared with that of the enemy. A Russian general, commanding the most solidly organized troops in Europe, need not fear to undertake any thing against undisciplined and unorganized troops in an open country, however brave may be its individuals.[1] Concert in action makes strength; order produces this concert, and discipline insures order; and without discipline and order no success is possible. The Russian general would not be so bold before European troops having the same instruction and nearly the same discipline as his own. Finally, a general may attempt with a Mack as his antagonist what it would be madness to do with a Napoleon.

The action of a cabinet in reference to the control of armies influences the boldness of their operations. A general whose genius and hands are tied by an Aulic council five hundred miles distant cannot be a match for one who has liberty of action, other things being equal.

As to superiority in skill, it is one of the most certain pledges of victory, all other things being equal. It is true that great generals have often been beaten by inferior ones; but an exception does not make a rule. An order misunderstood, a fortuitous event, may throw into the hands of the enemy all the chances of success which a skillful general had prepared for himself by his maneuvers. But these are risks which cannot be foreseen nor avoided. Would it be fair on that account to deny the influence of science and principles in ordinary affairs? This risk even proves the triumph of the principles, for it happens that they are applied accidentally by the army against which it was intended to apply them, and are the cause of its success. But, in admitting this truth, it may be said that it is an argument against science; this objection is not well founded, for a general's science consists in providing for his side all the chances possible to be foreseen, and of course cannot extend to the caprices of destiny. Even if the number of battles gained by skillful maneuvers did not exceed the number due to accident, it would not invalidate my assertion.

If the skill of a general is one of the surest elements of victory, it will readily be seen that the judicious selection of generals is one of the most delicate points in the science of government and one of the most essential parts of the military policy of a state. Unfortunately, this choice is influenced by so many petty passions, that chance, rank, age, favor, party spirit, jealousy, will

---

[1]   Irregular troops supported by disciplined troops may be of the greatest value, in destroying convoys, intercepting communication, &c., and may—as in the case of the French in 1812—make a retreat very disastrous

have as much to do with it as the public interest and justice. This subject is so important that we will devote to it a separate article.

ARTICLE XIII
MILITARY INSTITUTIONS

One of the most important points of the military policy of a state is the nature of its military institutions. A good army commanded by a general of ordinary capacity may accomplish great feats; a bad army with a good general may do equally well; but an army will certainly do a great deal more if its own superiority and that of the general be combined.

Twelve essential conditions concur in making a perfect army:–

1. To have a good recruiting-system;

2. A good organization;

8. A well-organized system of national reserves;

4. Good instruction of officers and men in drill and internal duties as well as those of a campaign;

5. A strict but not humiliating discipline, and a spirit of subordination and punctuality, based on conviction rather than on the formalities of the service;

6. A well-digested system of rewards, suitable to excite emulation;

7. The special arms of engineering and artillery to be well instructed;

8. An armament superior, if possible, to that of the enemy, both as to defensive and offensive arms;

9. A general staff capable of applying these elements, and having an organization calculated to advance the theoretical and practical education of its officers;

10. A good system for the commissariat, hospitals, and of general administration;

11. A good system of assignment to command, and of directing the principal operations of war;

12. Exciting and keeping alive the military spirit of the people.

To these conditions might be added a good system of clothing and equipment; for, if this be of less direct importance on the field of battle, it nevertheless has a bearing upon the preservation of the troops; and it is always a great object to economize the lives and health of veterans.

None of the above twelve conditions can be neglected without grave inconvenience. A fine army, well drilled and disciplined, but without national reserves, and unskillfully led, suffered Prussia to fall in fifteen days under the attacks of Napoleon. On the other hand, it has often been seen of how much advantage it is for a state to have a good army. It was the care and skill of Philip and Alexander in forming and instructing their phalanxes and rendering them easy to move, and capable of the most rapid maneuvers, which enabled the Macedonians to subjugate India and Persia with a handful of choice troops. It was the excessive love of his father for soldiers which procured for Frederick the Great an army capable of executing his great enterprises.

A government which neglects its army under any pretext whatever is thus culpable in the eyes of posterity, since it prepares humiliation for its standards and its country, instead of by a different course preparing for it success. We are far from saying that a government should sacrifice every thing to the army, for this would be absurd; but it ought to make the army the object of its constant care; and if the prince has not a military education it will be very difficult for him to fulfill his duty in this respect. In this case—which is, unfortunately, of too frequent occurrence—the defect must be supplied by wise institutions, at the head of which are to be placed a good system of the general staff, a good system of recruiting, and a good system of national reserves.

There are, indeed, forms of government which do not always allow the executive the power of adopting the best systems. If the armies of the Roman and French republics, and those of Louis XIV. and Frederick of Prussia, prove that a good military system and a skillful direction of operations may be found in governments the most opposite in principle, it cannot be doubted that, in the present state of the world, the form of government exercises a great influence in the development of the military strength of a nation and the value of its troops.

When the control of the public funds is in the hands of those affected by local interest or party spirit, they may be so over-scrupulous and penurious as to take all power to carry on the war from the executive, whom very many people seem to regard as a public enemy rather than as a chief devoted to all the national interests.

The abuse of badly-understood public liberties may also contribute to this deplorable result. Then it will be impossible for the most far-sighted administration to prepare in advance for a great war, whether it be demanded by the most important interests of the country at some future time, or whether it be immediate and necessary to resist sudden aggressions.

In the futile hope of rendering themselves popular, may not the members of an elective legislature, the majority of whom cannot be Richelieus, Pitts, or Louvois, in a misconceived spirit of economy, allow the institutions necessary for a large, well-appointed, and disciplined army to fall into decay? Deceived by the seductive fallacies of an exaggerated philanthropy, may they not end in convincing themselves and their constituents that the pleasures of peace are always preferable to the more statesmanlike preparations for war?

I am far from advising that states should always have the hand upon the sword and always be established on a war-footing: such a condition of things would be a scourge for the human race, and would not be possible, except under conditions not existing in all countries. I simply mean that civilized governments ought always to be ready to carry on a war in a short time,—that they should never be found unprepared. And the wisdom of their institutions may do as much in this work of preparation as foresight in their administration and the perfection of their system of military policy.

If, in ordinary times, under the rule of constitutional forms, governments subjected to all the changes of an elective legislature are less suitable than others for the creation or preparation of a formidable military power, nevertheless, in great crises these deliberative bodies have sometimes attained very different results, and have concurred in developing to the full extent the national strength. Still, the small number of such instances in history makes rather a list of exceptional cases, in which a tumultuous and violent assembly, placed under the necessity of conquering or perishing, has profited by the extraordinary enthusiasm of the nation to save the country and themselves at the same time by resorting to the most terrible measures and by calling to its aid an unlimited dictatorial power, which overthrew both liberty and law under the pretext of defending them. Here it is the dictatorship, or the absolute and monstrous usurpation of power, rather than the form of the deliberative assembly, which is the true cause of the display of energy. What happened in the Convention after the fall of Robespierre and the terrible Committee of Public Safety proves this, as well as the Chambers of 1815. Now, if the dictatorial power, placed in the hands of a few, has always been a plank of safety in great crises, it seems natural to draw the conclusion that countries controlled by elective assemblies must be politically and militarily weaker than pure monarchies, although in other respects they present decided advantages.

It is particularly necessary to watch over the preservation of armies in the interval of a long peace, for then they are most likely to degenerate. It is important to foster the military spirit in the armies, and to exercise them in great maneuvers, which, though but faintly resembling those of actual war, still are of decided advantage in preparing them for war. It is not less important to prevent them from becoming effeminate, which may be done by employing them in labors useful for the defense of the country.

The isolation in garrisons of troops by regiments is one of the worst possible systems, and the Russian and Prussian system of divisions and permanent corps d'armee seems to be much preferable. In general terms, the Russian army now may be presented as a model in many respects; and if in many points its customs would be useless and impracticable elsewhere, it must be admitted that many good institutions might well be copied from it.

As to rewards and promotion, it is essential to respect long service, and at the same time to open a way for merit. Three-fourths of the promotions in each grade should be made according to the roster, and the remaining fourth reserved for those distinguished for merit and zeal. On the contrary, in time of war the regular order of promotion should be suspended, or at least reduced to a third of the promotions, leaving the other two-thirds for brilliant conduct and marked services.

The superiority of armament may increase the chances of success in war: it does not, of itself, gain battles, but it is a great element of success. Every one can recall how nearly fatal to the French at Bylau and Marengo was their great inferiority in artillery. We may also refer to the great gain of the heavy French cavalry in the resumption of the cuirass, which they had for so long thrown aside. Every one knows the great advantage of the lance. Doubtless, as skirmishers lancers would not be more effectual than hussars, but when charging in line it is a very different affair. How many brave cavalry soldiers have been the victims of the prejudice they bore against the lance because it was a little more trouble to carry than a saber!

The armament of armies is still susceptible of great improvements; the state which shall take the lead in making them will secure great advantages. There is little left to be desired in artillery; but the offensive and defensive arms of infantry and cavalry deserve the attention of a provident government.

The new inventions of the last twenty years seem to threaten a great revolution in army organization, armament, and tactics. Strategy alone will remain unaltered, with its principles the same as under the Scipios and Caesars, Frederick and Napoleon, since they are independent of the nature of the arms and the organization of the troops.

The means of destruction are approaching perfection with frightful rapidity. [1] The Congreve rockets, the effect and direction of which it is said the Austrians can now regulate,—the shrapnel howitzers, which throw a stream of canister as far as the range of a bullet,—the Perkins steam-guns, which vomit forth as many balls as a battalion,—will multiply the chances of destruction, as though the hecatombs of Eylau, Borodino, Leipsic, and Waterloo were not sufficient to decimate the European races.

If governments do not combine in a congress to proscribe these inventions of destruction, there will be no course left but to make the half of an army consist of cavalry with cuirasses, in order to capture with great rapidity these machines; and the infantry, even, will be obliged to resume its armor of the Middle Ages, without which a battalion will be destroyed before engaging the enemy.

We may then see again the famous men-at-arms all covered with armor, and horses also will require the same protection.

While there is doubt about the realization of these fears, it is, however, certain that artillery and pyrotechny have made advances which should lead

---

[1]   It will be recollected that the author wrote this many years ago, since which time the inventive genius of the age has been attentively directed to the improvement of fire-arms. Artillery, which he regarded as almost perfect, has certainly undergone important improvements, and the improved efficiency of small arms is no less marked, while we hear nothing now of Perkins's steam-guns; and as yet no civilized army has been organized upon the plan the author suggests for depriving these destructive machines of their efficiency.—TRANSLATORS.

us to think of modifying the deep formation so much abused by Napoleon. We will recur to this in the chapter on Tactics.

We will here recapitulate, in a few words, the essential bases of the military policy which ought to be adopted by a wise government.

1. The prince should receive an education both political and military. He will more probably find men of administrative ability in his councils than good statesmen or soldiers; and hence he should be both of the latter himself.

2. If the prince in person does not lead his armies, it will be his first duty and his nearest interest to have his place well supplied. He must confide the glory of his reign and the safety of his states to the general most capable of directing his armies.

3. The permanent army should not only always be upon a respectable footing, but it should be capable of being doubled, if necessary, by reserves, which should always be prepared. Its instruction and discipline should be of a high character, as well as its organization; its armament should at least be as good as that of its neighbors, and superior if possible.

4. The materiel of war should also be upon the best footing, and abundant. The reserves should be stored in the depots and arsenals. National jealousy should not be allowed to prevent the adoption of all improvements in this materiel made in other countries.

5. It is necessary that the study of the military sciences should be encouraged and rewarded, as well as courage and zeal. The scientific military corps should be esteemed and honored: this is the only way of securing for the army men of merit and genius.

6. The general staff in times of peace should be employed in labors preparatory for all possible contingencies of war. Its archives should be furnished with numerous historical details of the past, and with all statistical, geographical, topographical, and strategic treatises and papers for the present and future. Hence it is essential that the chief of this corps, with a number of its officers, should be permanently stationed at the capital in time of peace, and the war-office should be simply that of the general staff, except that there should be a secret department for those documents to be concealed from the subalterns of the corps.

7. Nothing should be neglected to acquire a knowledge of the geography and the military statistics of other states, so as to know their material and moral capacity for attack and defense, as well as the strategic advantages of the two parties. Distinguished officers should be employed in these scientific labors, and should be rewarded when they acquit themselves with marked ability.

8. When a war is decided upon, it becomes necessary to prepare, not an entire plan of operations,—which is always impossible,—but a system of operations in reference to a prescribed aim; to provide a base, as well as all the material means necessary to guarantee the success of the enterprise.

9. The system of operations ought to be determined by the object of the war, the kind of forces of the enemy, the nature and resources of the country, the characters of the nations and of their chiefs, whether of the army or of the state. In fine, it should be based upon the moral and material means of attack or defense which the enemy may be able to bring into action; and it ought to take into consideration the probable alliances that may obtain in favor of or against either of the parties during the war.

10. The financial condition of a nation is to be weighed among the chances of a war. Still, it would be dangerous to constantly attribute to this condition the importance attached to it by Frederick the Great in the history of his times. He was probably right at his epoch, when armies were chiefly recruited by voluntary enlistment, when the last crown brought the last soldier; but when national levies are well organised money will no longer exercise the same influence,—at least for one or two campaigns. If England has proved that money will procure soldiers and auxiliaries, France has proved that love of country and honor are equally productive, and that, when necessary, war may be made to support war. France, indeed, in the fertility of her soil and the enthusiasm of her leaders, possessed sources of temporary power which cannot be adopted as a general base of a system; but the results of its efforts were none the less striking. Every year the numerous reports of the cabinet of London, and particularly of M. d'Yvernois, announced that France was about to break down for want of money, while Napoleon had 200,000,000 francs[1] in the vaults of the Tuileries, all the while meeting the expenses of the government, including the pay of his armies.

A power might be overrunning with gold and still defend itself very badly. History, indeed, proves that the richest nation is neither the strongest nor the happiest. Iron weighs at least as much as gold in the scales of military strength. Still, we must admit that a happy combination of wise military institutions, of patriotism, of well-regulated finances, of internal wealth and public credit, imparts to a nation the greatest strength and makes it best capable of sustaining a long war.

A volume would be necessary to discuss all the circumstances under which a nation may develop more or less strength, either by its gold or iron, and to determine the cases when war may be expected to support war. This result can only be obtained by carrying the army into the territory of the enemy; and all countries are not equally capable of furnishing resources to an assailant.

We need not extend further the investigation of these subjects which are not directly connected with the art of war. It is sufficient for our purpose to indicate their relations to a projected war; and it will be for the statesman to develop the modifications which circumstances and localities may make in these relations.

---

[1] There was a deficit in the finances of France at the fall of Napoleon. It was the result of his disasters, and of the stupendous efforts he was obliged to make. There was no deficit in 1811.

ARTICLE XIV

THE COMMAND OF ARMIES, AND THE CHIEF CONTROL OVER OPERATIONS

Is it an advantage to a state to have its armies commanded in person by the monarch? Whatever may be the decision on this point, it is certain that if the prince possess the genius of Frederick, Peter the Great, or Napoleon, he will be far from leaving to his generals the honor of performing great actions which he might do himself; for in this he would be untrue to his own glory and to the well-being of the country.

As it is not our mission to discuss the question whether it is more fortunate for a nation to have a warlike or a peace-loving prince, (which is a philanthropic question, foreign to our subject,) we will only state upon this point that, with equal merit and chances in other respects, a sovereign will always have an advantage over a general who is himself not the head of a state. Leaving out of the question that he is responsible only to himself for his bold enterprises, he may do much by the certainty he has of being able to dispose of all the public resources for the attainment of his end. He also possesses the powerful accessory of his favor, of recompenses and punishments; all will be devoted to the execution of his orders, and to insure for his enterprises the greatest success; no jealousy will interfere with the execution of his projects, or at least its exhibition will be rare and in secondary operations. Here are, certainly, sufficient motives to induce a prince to lead his armies, if he possess military capacity and the contest be of a magnitude worthy of him. But if he possess no military ability, if his character be feeble, and he be easily influenced, his presence with the army, instead of producing good results, will open the way for all manner of intrigues. Each one will present his projects to him; and, as he will not have the experience necessary to estimate them according to their merits, he will submit his judgment to that of his intimates. His general, interfered with and opposed in all his enterprises, will be unable to achieve success, even if he have the requisite ability. It may be said that a sovereign might accompany the army and not interfere with his general, but, on the contrary, aid him with all the weight of his influence. In this case his presence might be productive of good results, but it also might lead to great embarrassment. If the army were turned and cut off from its communications, and obliged to extricate itself, sword in hand, what sad results might not follow from the presence of the sovereign at head-quarters!

When a prince feels the necessity of taking the field at the head of his armies, but lacks the necessary self-confidence to assume the supreme direction of affairs, the best course will be that adopted by the Prussian government with Bluecher,—viz.; he should be accompanied by two generals of the best capacity, one of them a man of executive ability, the other a well-instructed staff officer. If this trinity be harmonious, it may yield excellent results, as in the case of the army of Silesia in 1813.

The same system might apply in the case where the sovereign judges it proper to intrust the command to a prince of his house, as has frequently happened since the time of Louis XIV. It has often occurred that the prince possessed only the titular command, and that an adviser, who in reality commanded, was imposed upon him. This was the case with the Duke of Orleans and Marsin at the famous battle of Turin, afterward with the Duke of Burgundy and Vendome at the battle of Audenarde, and, I think, also at Ulm with the Archduke Ferdinand and Mack. This system is deplorable, since no one is responsible for what is done. It is known that at the battle of Turin the Duke of Orleans exhibited more sagacity than Marsin, and it became necessary for the latter to show full secret authority from the king before the prince would yield his judgment and allow the battle to be lost. So at Ulm the archduke displayed more skill and courage than Mack, who was to be his mentor.

If the prince possess the genius and experience of the Archduke Charles, he should be invested with the untrammeled command, and be allowed full selection of his instruments. If he have not yet acquired the same titles to command, he may then be provided with an educated general of the staff, and another general distinguished for his talent in execution; but in no case will it be wise to invest either of these counselors with more authority than a voice in consultation.

We have already said that if the prince do not conduct his armies in person, his most important duty will be to have the position of commander well filled,— which, unfortunately, is not always done. Without going back to ancient times, it will be sufficient to recall the more modern examples under Louis XIV. and Louis XV. The merit of Prince Eugene was estimated by his deformed figure, and this drove him (the ablest commander of his time) into the ranks of the enemy. After Louvois' death, Tallard, Marsin, and Villeroi filled the places of Turenne, Conde, and Luxembourg, and subsequently Soubise and Clermont succeeded Marshal Saxe. Between the fashionable selections made in the Saloons of the Pompadours and Dubarrys, and Napoleon's preference for mere soldiers, there are many gradations, and the margin is wide enough to afford the least intelligent government means of making rational nominations; but, in all ages, human weaknesses will exercise an influence in one way or another, and artifice will often carry off the prize from modest or timid merit, which awaits a call for its services. But, leaving out of consideration all these influences, it will be profitable to inquire in what respects this choice of a commander will be difficult, even when the executive shall be most anxious to make it a judicious one. In the first place, to make choice of a skillful general requires either that the person who makes the selection shall be a military man, able to form an intelligent opinion, or that he should be guided by the opinions of others, which opens the way to the improper influence of cliques. The embarrassment is certainly less when there is at hand a general already illustrious by many victories; but, outside of the fact that every general is not a great leader because he has gained a battle, (for instance, Jourdan, Scherer, and many others,) it is not always the case that a victorious general is at the disposition of the government. It may well happen that after a long period of peace, there may not be a single general in Europe who has commanded in chief. In this case, it will be difficult to decide whether one general is better than another. Those who have served long in peace will be at the head of their arms or corps, and will have the rank appropriate for this position; but will they always be the most capable of filling it? Moreover, the intercourse of the heads of a government with their subordinates is generally so rare and transient, that it is not astonishing they should experience difficulty in assigning men to their appropriate positions. The judgment of the prince, misled by appearances, may err, and, with the purest intentions, he may well be deceived in his selections.

One of the surest means of escaping this misfortune would seem to be in realizing the beautiful fiction of Fenelon in Telemachus, by finding a faithful, sincere, and generous Philocles, who, standing between the prince and all aspirants for the command, would be able, by means of his more direct relations to the public, to enlighten the monarch in reference to selections of individuals best recommended by their character and abilities. But will this faithful friend never yield to personal affections? Will he be always free from prejudice? Suwaroff was rejected by Potemkin on account of his appearance, and it required all the art of Catherine to secure a regiment for the man who afterward shed so much luster upon the Russian arms.

It has been thought that public opinion is the best guide; but nothing could be more dangerous. It voted Dumouriez to be a Caesar, when he was ignorant of the great operations of war. Would it have placed Bonaparte at the head of the army of Italy, when he was known only by two directors? Still, it must be admitted that, if not infallible, public sentiment is not to be despised, particularly if it survive great crises and the experience of events.

The most essential qualities for a general will always be as follow:–First, *A high moral courage, capable of great resolutions*; Secondly, *A physical courage which takes no account of danger*. His scientific or military acquirements are secondary to the above-mentioned characteristics, though if great they will be valuable auxiliaries. It is not necessary that he should be a man of vast erudition. His knowledge may be limited, but it should be thorough, and he should be perfectly grounded in the principles at the base of the art of war. Next in importance come the qualities of his personal character. A man who is gallant, just, firm, upright, capable of esteeming merit in others instead of being jealous of it, and skillful in making this merit conduce to his own glory, will always be a good general, and may even pass for a great man. Unfortunately, the disposition to do justice to merit in others is not the most common quality: mediocre minds are always jealous, and inclined to surround themselves with persons of little ability, fearing the reputation of being led, and not realizing that the nominal commander of an army always receives almost all the glory of its success, even when least entitled to it.

The question has often been discussed, whether it is preferable to assign to the command a general of long experience in service with troops, or an officer of the staff, having generally but little experience in the management of troops. It is beyond question that war is a distinct science of itself, and that it is quite possible to be able to combine operations skillfully without ever having led a regiment against an enemy. Peter the Great, Conde, Frederick, and Napoleon are instances of it. It cannot, then, be denied that an officer from the staff may as well as any other prove to be a great general, but it will not be because he has grown gray in the duties of a quartermaster that he will be capable of the supreme command, but because he has a natural genius for war and possesses the requisite characteristics. So, also, a general from the ranks of the infantry or cavalry may be as capable of conducting a campaign as the most profound tactician. So this question does not admit of a definite answer either in the affirmative or negative, since almost all will depend upon the personal qualities of the individuals; but the following remarks will be useful in leading to a rational conclusion:–

1. A general, selected from the general staff, engineers, or artillery, who has commanded a division or a corps d'armee, will, with equal chances, be superior to one who is familiar with the service of but one arm or special corps.

2. A general from the line, who has made a study of the science of war, will be equally fitted for the command.

3. That the character of the man is above all other requisites in a commander-in-chief.

Finally, He will be a good general in whom are found united the requisite personal characteristics and a thorough knowledge of the principles of the art of war.

The difficulty of always selecting a good general has led to the formation of a good general staff, which being near the general may advise him, and thus exercise a beneficial influence over the operations. A well-instructed general staff is one of the most useful of organizations; but care must be observed to prevent the introduction into it of false principles, as in this case it might prove fatal.

Frederick, when he established the military school of Potsdam, never thought it would lead to the "right shoulder forward" of General Ruchel,[1] and to the teaching that the oblique order is the infallible rule for gaining all battles. How true it is that there is but a step from the sublime to the ridiculous!

Moreover, there ought to exist perfect harmony between the general and his chief of staff; and, if it be true that the latter should be a man of recognized ability, it is also proper to give the general the choice of the men who are to be his advisers. To impose a chief of staff upon a general would be to create anarchy and want of harmony; while to permit him to select a cipher for that position would be still more dangerous; for if he be himself a man of little

---

[1]   General Ruchel thought at the battle of Jena that he could save the army by giving the command to advance the right shoulder in order to form an oblique line.

ability, indebted to favor or fortune for his station, the selection will be of vital importance. The best means to avoid these dangers is to give the general the option of several designated officers, all of undoubted ability.

It has been thought, in succession, in almost all armies, that frequent councils of war, by aiding the commander with their advice, give more weight and effect to the direction of military operations. Doubtless, if the commander were a Soubise, a Clermont, or a Mack, he might well find in a council of war opinions more valuable than his own; the majority of the opinions given might be preferable to his; but what success could be expected from operations conducted by others than those who have originated and arranged them? What must be the result of an operation which is but partially understood by the commander, since it is not his own conception?

I have undergone a pitiable experience as prompter at head-quarters, and no one has a better appreciation of the value of such services than myself; and it is particularly in a council of war that such a part is absurd. The greater the number and the higher the rank of the military officers who compose the council, the more difficult will it be to accomplish the triumph of truth and reason, however small be the amount of dissent.

What would have been the action of a council of war to which Napoleon proposed the movement of Arcola, the crossing of the Saint-Bernard, the maneuver at Ulm, or that at Gera and Jena? The timid would have regarded them as rash, even to madness, others would have seen a thousand difficulties of execution, and all would have concurred in rejecting them; and if, on the contrary, they had been adopted, and had been executed by any one but Napoleon, would they not certainly have proved failures?

In my opinion, councils of war are a deplorable resource, and can be useful only when concurring in opinion with the commander, in which case they may give him more confidence in his own judgment, and, in addition, may assure him that his lieutenants, being of his opinion, will use every means to insure the success of the movement. This is the only advantage of a council of war, which, moreover, should be simply consultative and have no further authority; but if, instead of this harmony, there should be difference of opinion, it can only produce unfortunate results.

Accordingly, I think it safe to conclude that the best means of organizing the command of an army, in default of a general approved by experience, is—

1st. To give the command to a man of tried bravery, bold in the fight, and of unshaken firmness in danger.

2d. To assign, as his chief of staff, a man of high ability, of open and faithful character, between whom and the commander there may be perfect harmony. The victor will gain so much glory that he can spare some to the friend who has contributed to his success. In this way Bluecher, aided by Gneisenau and Muffling, gained glory which probably he would not have been able to do of himself. It is true that this double command is more objectionable than an undivided one when a state has a Napoleon, a Frederick, or a Suwaroff to fill it; but when there is no great general to lead the armies it is certainly the preferable system.

Before leaving this important branch of the subject, another means of influencing military operations—viz.: that of a council of war at the seat of government—deserves notice. Louvois for a long time directed from Paris the armies of Louis XIV., and with success. Carnot, also, from Paris directed the armies of the Republic: in 1793 he did well, and saved France; in 1794 his action was at first very unfortunate, but he repaired his faults afterward by chance; in 1796 he was completely at fault. It is to be observed, however, that both Louvois and Carnot individually controlled the armies, and that there was no council of war. The Aulic council, sitting in Vienna, was often intrusted with the duty of directing the operations of the armies; and there has never been but one opinion in Europe as to its fatal influence. Whether this opinion is right or wrong, the Austrian generals alone are able to decide. My own opinion is that the functions of such a body in this connection should be limited to the adoption of a general plan of operations. By this I do not mean a plan which should trace out the campaign in detail, restricting the generals and compelling them to give battle without regard to circumstances, but a plan which should determine the object of the campaign, the nature of the operations, whether offensive or defensive, the material means to be applied to these first enterprises, afterward for the reserves, and finally for the levies which may be necessary if the country be invaded. These points, it is true, should be discussed in a council of both generals and ministers, and to these points should the control of the council be limited; for if it should not only order the general in command to march to Vienna or to Paris, but should also have the presumption to indicate the manner in which he should maneuver to attain this object, the unfortunate general would certainly be beaten, and the whole responsibility of his reverses should fall upon the shoulders of those who, hundreds of miles distant, took upon themselves the duty of directing the army,—a duty so difficult for any one, even upon the scene of operations.

ARTICLE XV

THE MILITARY SPIRIT OF NATIONS, AND THE MORALE OF ARMIES

The adoption of the best regulations for the organization of an army would be in vain if the government did not at the same time cultivate a military spirit in its citizens. It may well be the case in London, situated on an island and protected from invasion by its immense fleets, that the title of a rich banker should be preferred to a military decoration; but a continental nation imbued with the sentiments and habits of the tradesmen of London or the bankers of Paris would sooner or later fall a prey to its neighbors. It was to the union of the civic virtues and military spirit fostered by their institutions that the

Romans were indebted for their grandeur; and when they lost these virtues, and when, no longer regarding the military service as an honor as well as a duty, they relinquished it to mercenary Goths and Gauls, the fall of the empire became inevitable. It is doubtless true that whatever increases the prosperity of the country should be neither neglected nor despised; it is also necessary to honor the branches of industry which are the first instruments of this prosperity; but they should always be secondary to the great institutions which make up the strength of states in encouraging the cultivation of the manly and heroic virtues. Policy and justice both agree on this point; for, whatever Boileau may say, it is certainly more glorious to confront death in the footsteps of the Caesars than to fatten upon the public miseries by gambling on the vicissitudes of the national credit. Misfortune will certainly fall upon the land where the wealth of the tax-gatherer or the greedy gambler in stocks stands, in public estimation, above the uniform of the brave man who sacrifices his life, health, or fortune to the defense of his country.

The first means of encouraging the military spirit is to invest the army with all possible social and public consideration. The second means is to give the preference to those who have rendered services to the state, in filling any vacancies in the administrative departments of the government, or even to require a certain length of military service as a qualification for certain offices. A comparison of the ancient military institutions of Rome with those of Russia and Prussia, is a subject worthy of serious attention; and it would also be interesting to contrast them with the doctrines of modern theorists, who declare against the employment of officers of the army in other public functions, and who wish for none but rhetoricians in the important offices of administration.[1] It is true that many public employments demand a special course of study; but cannot the soldier, in the abundant leisure of peace, prepare himself for the career he would prefer after having fulfilled his debt to his country in the profession of arms? If these administrative offices were conferred upon officers retired from the army in a grade not lower than that of captain, would it not be a stimulant for officers to attain that rank, and would it not lead them, when in garrisons, to find their recreations elsewhere than in the theaters and public clubs?

It may be possible that this facility of transfer from the military to the civil service would be rather injurious than favorable to a high military spirit, and that to encourage this spirit it would be expedient to place the profession of the soldier above all others. This was the early practice of the Mamelukes and Janissaries. Their soldiers were bought at the age of about seven years, and were educated in the idea that they were to die by their standards. Even the English—so jealous of their rights—contract, in enlisting as soldiers, the obligation for the whole length of their lives, and the Russian, in enlisting for twenty-five years, does what is almost equivalent. In such armies, and in

---

[1] For instance, in France, instead of excluding all officers from the privilege of the elective franchise, it should be given to all colonels; and the generals should be eligible to the legislature. The most venal deputies will not be those from military life.

those recruited by voluntary enlistments, perhaps it would not be advisable to tolerate this fusion of military and civil offices; but where the military service is a temporary duty imposed upon the people, the case is different, and the old Roman laws which required a previous military service of ten years in any aspirant for the public employments, seem to be best calculated to preserve the military spirit,—particularly in this age, when the attainment of material comfort and prosperity appears to be the dominant passion of the people.

However this may be, still, in my opinion, under all forms of government, it will be a wise part to honor the military profession, in order to encourage the love of glory and all the warlike virtues, under the penalty of receiving the reproaches of posterity and suffering insult and dependency.

It is not sufficient to foster the military spirit among the people, but, more than that, it is necessary to encourage it in the army. Of what avail would it be if the uniform be honored in the land and it be regarded as a duty to serve in the army, while the military virtues are wanting? The forces would be numerous but without valor.

The enthusiasm of an army and its military spirit are two quite different things, and should not be confounded, although they produce the same effects. The first is the effect of passions more or less of a temporary character,—of a political or religious nature, for instance, or of a great love of country; while the latter, depending upon the skill of the commander and resulting from military institutions, is more permanent and depends less upon circumstances, and should be the object of the attention of every far-seeing government.[1] Courage should be recompensed and honored, the different grades in rank respected, and discipline should exist in the sentiments and convictions rather than in external forms only.

The officers should feel the conviction that resignation, bravery, and faithful attention to duty are virtues without which no glory is possible, no army is respectable, and that firmness amid reverses is more honorable than enthusiasm in success,—since courage alone is necessary to storm a position, while it requires heroism to make a difficult retreat before a victorious and enterprising enemy, always opposing to him a firm and unbroken front. A fine retreat should meet with a reward equal to that given for a great victory.

By inuring armies to labor and fatigue, by keeping them from stagnation in garrison in times of peace, by inculcating their superiority over their enemies, without depreciating too much the latter, by inspiring a love for great exploits,—in a word, by exciting their enthusiasm by every means in harmony with their tone of mind, by honoring courage, punishing weakness, and disgracing cowardice,—we may expect to maintain a high military spirit.

Effeminacy was the chief cause of the ruin of the Roman legions: those formidable soldiers, who had borne the casque, buckler, and cuirass in the

---

[1]   It is particularly important that this spirit should pervade the officers and non-commissioned officers: if they be capable, and the nation brave, there need be no fear for the men.

times of the Scipios under the burning sun of Africa, found them too heavy in the cool climates of Germany and Gaul; and then the empire was lost.

I have remarked that it is not well to create a too great contempt for the enemy, lest the *morale* of the soldier should be shaken if he encounter an obstinate resistance. Napoleon at Jena, addressing Lannes' troops, praised the Prussian cavalry, but promised that they would contend in vain against the bayonets of his Egyptians.

The officers and troops must be warned against those sudden panics which often seize the bravest armies when they are not well controlled by discipline, and hence when they do not recognize that in order is the surest hope of safety. It was not from want of courage that one hundred thousand Turks were beaten at Peterwardein by Prince Eugene, and at Kagoul by Romanzoff: it was because, once repulsed in their disorderly charges, every one yielded to his personal feelings, and because they fought individually, but not in masses and in order. An army seized with panic is similarly in a state of demoralization; because when disorder is once introduced all concerted action on the part of individuals becomes impossible, the voice of the officers can no longer be heard, no maneuver for resuming the battle can be executed, and there is no resource but in ignominious flight.

Nations with powerful imaginations are particularly liable to panics; and nothing short of strong institutions and skillful leaders can remedy it. Even the French, whose military virtues when well led have never been questioned, have often performed some quick movements of this kind which were highly ridiculous. We may refer to the unbecoming panic which pervaded the infantry of Marshal Villars after having gained the battle of Friedlingen, in 1704. The same occurred to Napoleon's infantry after the victory of Wagram and when the enemy was in full retreat. A still more extraordinary case was the flight of the 97th semi-brigade, fifteen hundred strong, at the siege of Genoa, before a platoon of cavalry. Two days afterward these same men took Fort Diamond by one of the most vigorous assaults mentioned in modern history.

Still, it would seem to be easy to convince brave men that death comes more quickly and more surely to those who fly in disorder than to those who remain together and present a firm front to the enemy, or who rally promptly when their lines have been for the instant broken.

In this respect the Russian army may be taken as a model by all others. The firmness which it has displayed in all retreats is due in equal degrees to the national character, the natural instincts of the soldiers, and the excellent disciplinary institutions. Indeed, vivacity of imagination is not always the cause of the introduction of disorder: the want of the habit of order often causes it, and the lack of precautions on the part of the generals to maintain this order contributes to it. I have often been astonished at the indifference of most generals on this point. Not only did they not deign to take the slightest precaution to give the proper direction to small detachments or scattered men, and fail to adopt any signals to facilitate the rallying in each division of the fractions which may be scattered in a momentary panic or in an irresistible charge of the enemy, but they were offended that any one should think of proposing such precautions. Still, the most undoubted courage and the most severe discipline will often be powerless to remedy a great disorder, which might be in a great degree obviated by the use of rallying-signals for the different divisions. There are, it is true, cases where all human resources are insufficient for the maintenance of order, as when the physical sufferings of the soldiers have been so great as to render them deaf to all appeals, and when their officers find it impossible to do any thing to organize them,—which was the case in the retreat of 1812. Leaving out these exceptional cases, good habits of order, good logistical precautions for rallying, and good discipline will most frequently be successful, if not in preventing disorder, at least in promptly remedying it.

It is now time to leave this branch, of which I have only desired to trace an outline, and to proceed to the examination of subjects which are purely military.

COMMENTARY ON CHAPTER II

*T*he chapter on military police emphasizes the importance of good intelligence or, as Jomini puts it, military statistics. Accurate knowledge of the geography, demographics, and military strength of an opponent are obviously necessary in order to prosecute a war successfully, but such information has often been ignored by generals and policymakers. Napoleon III had no idea of the military strength of Prussia in 1870; more recently, Mussolini ignored the effect that the rugged Greek terrain would have on Italian operations in 1940. Other articles in Chapter 2 begin narrowing the emphasis of the book to more purely military subjects. Jomini discusses the form that national military institutions should take, the relationship between the head of state and the principal military commander, and the organization of military headquarters. Article XV is Jomini's nod to morale in this book, a subject he generally ignores in the chapters focusing more directly on military operations.

CHAPTER III

STRATEGY

DEFINITION OF STRATEGY AND THE FUNDAMENTAL PRINCIPLE OF WAR

The art of war, independently of its political and moral relations, consists of five principal parts, viz.: Strategy, Grand Tactics, Logistics, Tactics of the different arms, and the Art of the Engineer. We will treat of the first three branches, and begin by defining them. In order to do this, we will follow the order of procedure of a general when war is first declared, who commences with the points of the highest importance, as a plan of campaign, and afterward descends to the necessary details. Tactics, on the contrary, begins with details, and ascends to combinations and generalization necessary for the formation and handling of a great army.

We will suppose an army taking the field: the first care of its commander should be to agree with the head of the state upon the character of the war: then he must carefully study the theater of war, and select the most suitable base of operations, taking into consideration the frontiers of the state and those of its allies.

The selection of this base and the proposed aim will determine the zone of operations. The general will take a first objective point: he will select the line of operations leading to this point, either as a temporary or permanent line, giving it the most advantageous direction; namely, that which promises the greatest number of favorable opportunities with the least danger. An army marching on this line of operations will have a front of operations and a strategic front. The temporary positions which the corps d'armee will occupy upon this front of operations, or upon the line of defense, will be strategic positions.

When near its first objective point, and when it begins to meet resistance, the army will either attack the enemy or maneuver to compel him to retreat; and for this end it will adopt one or two strategic lines of maneuvers, which, being temporary, may deviate to a certain degree from the general line of operations, with which they must not be confounded.

To connect the strategic front with the base as the advance is made, lines of supply, depots, &c. will be established.

If the line of operations be long, and there be hostile troops in annoying proximity to it, these bodies may either be attacked and dispersed or be merely observed, or the operations against the enemy may be carried on without reference to them. If the second of these courses be pursued, a double strategic front and large detachments will be the result.

The army being almost within reach of the first objective point, if the enemy oppose him there will be a battle; if indecisive, the fight will be resumed; if the army gains the victory, it will secure its objective point or will advance to attain a second. Should the first objective point be the possession of an important fort, the siege will be commenced. If the army be not strong enough to continue its march, after detaching a sufficient force to maintain the siege, it will take a strategic position to cover it, as did the army of Italy in 1796, which, less than fifty thousand strong, could not pass Mantua to enter Austria, leaving twenty-five thousand enemies within its walls, and having forty thousand more in front on the double line of the Tyrol and Frioul.

If the army be strong enough to make the best use of its victory, or if it have no siege to make, it will operate toward a second and more important objective point.

If this point be distant, it will be necessary to establish an intermediate point of support. One or more secure cities already occupied will form an eventual base: when this cannot be done, a small strategic reserve may be established, which will protect the rear and also the depots by temporary fortifications. When the army crosses large streams, it will construct *tetes de pont*; and, if the bridges are within walled cities, earth-works will be thrown up to increase the means of defense and to secure the safety of the eventual base or the strategic reserve which may occupy these posts.

Should the battle be lost, the army will retreat toward its base, in order to be reinforced therefrom by detachments of troops, or, what is equivalent, to strengthen itself by the occupation of fortified posts and camps, thus compelling the enemy to halt or to divide his forces.

When winter approaches, the armies will either go into quarters, or the field will be kept by the army which has obtained decisive success and is desirous of profiting to the utmost by its superiority. These winter campaigns are very trying to both armies, but in other respects do not differ from ordinary campaigns, unless it be in demanding increased activity and energy to attain prompt success.

Such is the ordinary course of a war, and as such we will consider it, while discussing combinations which result from these operations.

Strategy embraces the following points, viz.:–

1. The selection of the theater of war, and the discussion of the different combinations of which it admits.

2. The determination of the decisive points in these combinations, and the most favorable direction for operations.

3. The selection and establishment of the fixed base and of the zone of operations.

4. The selection of the objective point, whether offensive or defensive.

5. The strategic fronts, lines of defense, and fronts of operations.

6. The choice of lines of operations leading to the objective point or strategic front.

7. For a given operation, the best strategic line, and the different maneuvers necessary to embrace all possible cases.

8. The eventual bases of operations and the strategic reserves.

9. The marches of armies, considered as maneuvers.

10. The relation between the position of depots and the marches of the army.

11. Fortresses regarded as strategical means, as a refuge for an army, as an obstacle to its progress: the sieges to be made and to be covered.

12. Points for intrenched camps, *tetes de pont*, &c.

13. The diversions to be made, and the large detachments necessary.

These points are principally of importance in the determination of the first steps of a campaign; but there are other operations of a mixed nature, such as passages of streams, retreats, surprises, disembarkations, convoys, winter quarters, the execution of which belongs to tactics, the conception and arrangement to strategy.

The maneuvering of an army upon the battle-field, and the different formations of troops for attack, constitute Grand Tactics. Logistics is the art of moving armies. It comprises the order and details of marches and camps, and of quartering and supplying troops; in a word, it is the execution of strategical and tactical enterprises.

To repeat. Strategy is the art of making war upon the map, and comprehends the whole theater of operations. Grand Tactics is the art of posting troops upon the battle-field according to the accidents of the ground, of bringing them into action, and the art of fighting upon the ground, in contradistinction to planning upon a map. Its operations may extend over a field of ten or twelve miles in extent. Logistics comprises the means and arrangements which work out the plans of strategy and tactics. Strategy decides where to act; logistics brings the troops to this point; grand tactics decides the manner of execution and the employment of the troops.

It is true that many battles have been decided by strategic movements, and have been, indeed, but a succession of them; but this only occurs in the exceptional case of a dispersed army: for the general case of pitched battles the above definition holds good.

Grand Tactics, in addition to acts of local execution, relates to the following objects:–

1. The choice of positions and defensive lines of battle.

2. The offensive in a defensive battle.

3. The different orders of battle, or the grand maneuvers proper for the attack of the enemy's line.

4. The collision of two armies on the march, or unexpected battles.

5. Surprises of armies in the open field.

6. The arrangements for leading troops into battle.

7. The attack of positions and intrenched camps.

8. *Coups de main.*

All other operations, such as relate to convoys, foraging-parties, skirmishes of advanced or rear guards, the attack of small posts, and any thing accomplished by a detachment or single division, may be regarded as details of war, and not included in the great operations.

THE FUNDAMENTAL PRINCIPLE OF WAR

It is proposed to show that there is one great principle underlying all the operations of war,—a principle which must be followed in all good combinations. It is embraced in the following maxims:–

1. To throw by strategic movements the mass of an army, successively, upon the decisive points of a theater of war, and also upon the communications of the enemy as much as possible without compromising one's own.

2. To maneuver to engage fractions of the hostile army with the bulk of one's forces.

3. On the battle-field, to throw the mass of the forces upon the decisive point, or upon that portion of the hostile line which it is of the first importance to overthrow.

4. To so arrange that these masses shall not only be thrown upon the decisive point, but that they shall engage at the proper times and with energy.

This principle has too much simplicity to escape criticism: one objection is that it is easy to recommend throwing the mass of the forces upon the decisive points, but that the difficulty lies in recognizing those points.

This truth is evident; and it would be little short of the ridiculous to enunciate such a general principle without accompanying it with all necessary explanations for its application upon the field. In Article XIX. these decisive points will be described, and in Articles from XVIII. to XXII. will be discussed their relations to the different combinations. Those students who, having attentively considered what is there stated, still regard the determination of these points as a problem without a solution, may well despair of ever comprehending strategy.

The general theater of operations seldom contains more than three zones,— the right, the left, and the center; and each zone, front of operations, strategic position, and line of defense, as well as each line of battle, has the same subdivisions,—two extremities and the center. A direction upon one of these three will always be suitable for the attainment of the desired end. A direction upon one of the two remaining will be less advantageous; while the third direction will be wholly inapplicable. In considering the object proposed in connection with the positions of the enemy and the geography of the country, it will appear that in every strategic movement or tactical maneuver the question for decision will always be, whether to maneuver to the right, to the left, or directly in front. The selection of one of these three simple alternatives cannot, surely, be considered an enigma. The art of giving the proper direction to the masses is certainly the basis of strategy, although it is not the whole of the art of war. Executive talent, skill, energy, and a quick apprehension of events are necessary to carry out any combinations previously arranged.

We will apply this great principle to the different cases of strategy and tactics, and then show, by the history of twenty celebrated campaigns, that, with few exceptions, the most brilliant successes and the greatest reverses resulted from an adherence to this principle in the one case, and from a neglect of it in the other.

OF STRATEGIC COMBINATIONS

ARTICLE XVI
OF THE SYSTEM OF OPERATIONS

War once determined upon, the first point to be decided is, whether it shall be offensive or defensive; and we will first explain what is meant by these terms. There are several phases of the offensive: if against a great state, the whole or a large portion of whose territory is attacked, it is an *invasion*; if a province only, or a line of defense of moderate extent, be assailed, it is the ordinary offensive; finally, if the offensive is but an attack upon the enemy's position, and is confined to a single operation, it is called the taking the *initiative*. In a moral and political view, the offensive is nearly always advantageous: it carries the war upon foreign soil, saves the assailant's country from devastation, increases his resources and diminishes those of his enemy, elevates the *morale* of his army, and generally depresses the adversary. It sometimes happens that invasion excites the ardor and energy of the adversary,—particularly when he feels that the independence of his country is threatened.

In a military point of view, the offensive has its good and its bad side. Strategically, an invasion leads to deep lines of operations, which are always dangerous in a hostile country. All the obstacles in the enemy's country, the mountains, rivers, defiles, and forts, are favorable for defense, while the inhabitants and authorities of the country, so far from being the instruments of the invading army, are generally hostile. However, if success be obtained, the enemy is struck in a vital point: he is deprived of his resources and compelled to seek a speedy termination of the contest.

For a single operation, which we have called the taking the *initiative*, the offensive is almost always advantageous, particularly in strategy. Indeed, if the art of war consists in throwing the masses upon the decisive points, to do this it will be necessary to take the initiative. The attacking party knows what he is doing and what he desires to do; he leads his masses to the point where he desires to strike. He who awaits the attack is everywhere anticipated: the enemy fall with large force upon fractions of his force: he neither knows where his adversary proposes to attack him nor in what manner to repel him.

Tactically, the offensive also possesses advantages, but they are less positive, since, the operations being upon a limited field, the party taking the initiative cannot conceal them from the enemy, who may detect his designs and by the aid of good reserves cause them to fail.

The attacking party labors under the disadvantages arising from the obstacles to be crossed before reaching the enemy's line; on which account the advantages and disadvantages of the tactical offensive are about equally balanced.

Whatever advantages may be expected either politically or strategically from the offensive, it may not be possible to maintain it exclusively throughout the war; for a campaign offensive in the beginning may become defensive before it ends.

A defensive war is not without its advantages, when wisely conducted. It may be passive or active, taking the offensive at times. The passive defense is always pernicious; the active may accomplish great successes. The object of a defensive war being to protect, as long as possible, the country threatened by the enemy, all operations should be designed to retard his progress, to annoy him in his enterprises by multiplying obstacles and difficulties, without, however, compromising one's own army. He who invades does so by reason of some superiority; he will then seek to make the issue as promptly as possible: the defense, on the contrary, desires delay till his adversary is weakened by sending off detachments, by marches, and by the privations and fatigues incident to his progress.

An army is reduced to the defensive only by reverses or by a positive inferiority. It then seeks in the support of forts, and in natural or artificial barriers, the means of restoring equality by multiplying obstacles in the way of the enemy. This plan, when not carried to an extreme, promises many chances of success, but only when the general has the good sense not to make the defense passive: he must not remain in his positions to receive whatever blows may be given by his adversary; he must, on the contrary, redouble his activity, and be constantly upon the alert to improve all opportunities of assailing the weak points of the enemy. This plan of war may be called the defensive-offensive, and may have strategical as well as tactical advantages.. It combines the advantages of both systems; for one who awaits his adversary upon a prepared field, with all his own resources in hand, surrounded by all the advantages of being on his own ground, can with hope of success take the initiative, and is fully able to judge when and where to strike.

During the first three campaigns of the Seven Years' War Frederick was the assailant; in the remaining four his conduct was a perfect model of the defensive-offensive. He was, however, wonderfully aided in this by his adversaries, who allowed him all the time he desired, and many opportunities of taking the offensive with success. Wellington's course was mainly the same in Portugal, Spain, and Belgium, and it was the most suitable in his circumstances. It seems plain that one of the greatest talents of a general is to know how to use (it may be alternately) these two systems, and particularly to be able to take the initiative during the progress of a defensive war.

ARTICLE XVII

OF THE THEATER OF OPERATIONS

The theater of a war comprises all the territory upon which the parties may assail each other, whether it belong to themselves, their allies, or to weaker states who may be drawn into the war through fear or interest. When the war is also maritime, the theater may embrace both hemispheres,—as has happened in contests between France and England since the time of Louis XIV. The theater of a war may thus be undefined, and must, not be confounded with the theater of operations of one or the other army. The theater of a continental war between France and Austria may be confined to Italy, or may, in addition, comprise Germany if the German States take part therein.

Armies may act in concert or separately: in the first case the whole theater of operations may be considered as a single field upon which strategy directs the armies for the attainment of a definite end. In the second case each army will have its own independent theater of operations. The *theater of operations* of an army embraces all the territory it may desire to invade and all that it may be necessary to defend. If the army operates independently, it should not attempt any maneuver beyond its own theater, (though it should leave it if it be in danger of being surrounded,) since the supposition is that no concert of action has been arranged with the armies operating on the other fields. If, on the contrary, there be concert of action, the theater of operations of each army taken singly is but a zone of operations of the general field, occupied by the masses for the attainment of a common object.

Independently of its topographical features, each theater upon which one or more armies operate is composed, for both parties, as follows:–

1. Of a fixed base of operations.
2. Of a principal objective point.
3. Of fronts of operations, strategic fronts, and lines of defense.
4. Of zones and lines of operations.
5. Of temporary strategic lines and lines of communications.
6. Of natural or artificial obstacles to be overcome or to oppose to the enemy.
7. Of geographical strategic points, whose occupation is important, either for the offensive or defensive.
8. Of accidental intermediate bases of operations between the objective point and the primary base.
9. Of points of refuge in case of reverse.

For illustration, let us suppose the case of France invading Austria with two or three armies, to be concentrated under one commander, and starting from Mayence, from the Upper Rhine, from Savoy or the Maritime Alps, respectively. The section of country which each of these armies traverses may be considered as a zone of the general field of operations. But if the army of Italy goes but to the Adige without concerted action with the army of the Rhine, then what was before but a zone becomes for that army a theater of operations.

In every case, each theater must have its own base, its own objective point, its zones and lines of operations connecting the objective point with the base, either in the offensive or the defensive.

It has been taught and published that rivers are lines of operations *par excellence*. Now, as such a line must possess two or three roads to move the army within the range of its operations, and at least one line of retreat, rivers have been called lines of retreat, and even lines of maneuver. It would be much more accurate to say that rivers are excellent lines of supply, and powerful auxiliaries in the establishment of a good line of operations, but never the line itself.

It has also been maintained that, could one create a country expressly to be a good theater of war, converging roads would be avoided, because they facilitate invasion. Every country has its capital, its rich cities for manufactures or trade; and, in the very nature of things, these points must be the centers of converging routes. Could Germany be made a desert, to be molded into a theater of war at the pleasure of an individual, commercial cities and centers of trade would spring up, and the roads would again necessarily converge to these points. Moreover, was not the Archduke Charles enabled to beat Jourdan in 1796 by the use of converging routes? Besides, these routes are more favorable for defense than attack, since two divisions retreating upon these radial lines can effect a junction more quickly than two armies which are pursuing, and they may thus united defeat each of the pursuing masses separately.

Some authors have affirmed that mountainous countries abound in strategic positions; others have maintained that, on the contrary, these points are more rare among the Alps than in the plains, but also that if more rare they are more important and more decisive.

Some authors have represented that high ranges of mountains are, in war, inaccessible barriers. Napoleon, on the contrary, in speaking of the Rhetian Alps, said that "an army could pass wherever a man could put his foot."

Generals no less experienced than himself in mountain-warfare have united with him in this opinion, in admitting the great difficulty of carrying on a defensive war in such localities unless the advantages of partisan and regular warfare can be combined, the first to guard the heights and to harass the enemy, the second to give battle at the decisive points,—the junctions of the large valleys.

These differences of opinion are here noticed merely to show the reader that, so far from the art having reached perfection, there are many points that admit of discussion.

The most important topographical or artificial features which make up the theater of a war will, in succeeding portions of this chapter, be examined as to their strategic value; but here it may be proper to remark that this value will depend much upon the spirit and skill of the general. The great leader who crossed the Saint-Bernard and ordered the passage of the Splugen was far from believing in the impregnability of these chains; but he was also far from thinking that a muddy rivulet and a walled inclosure could change his destiny at Waterloo.

ARTICLE XVIII
BASES OF OPERATIONS

A base of operations is the portion of country from which the army obtains its reinforcements and resources, from which it starts when it takes the offensive, to which it retreats when necessary, and by which it is supported when it takes position to cover the country defensively.

The base of operations is most generally that of supply,—though not necessarily so, at least as far as food is concerned; as, for instance, a French army upon the Elbe might be subsisted from Westphalia or Franconia, but its real base would certainly be upon the Rhine.

When a frontier possesses good natural or artificial barriers, it may be alternately either an excellent base for offensive operations, or a line of defense when the state is invaded. In the latter case it will always be prudent to have a second base in rear; for, although an army in its own country will everywhere find a point of support, there is still a vast difference between those parts of the country without military positions and means, as forts, arsenals, and fortified depots, and those other portions where these military resources are found; and these latter alone can be considered as safe bases of operations. An army may have in succession a number of bases: for instance, a French army in Germany will have the Rhine for its first base; it may have others beyond this, wherever it has allies or permanent lines of defense; but if it is driven back across the Rhine it will have for a base either the Meuse or the Moselle: it might have a third upon the Seine, and a fourth upon the Loire.

These successive bases may not be entirely or nearly parallel to the first. On the contrary, a total change of direction may become necessary. A French army repulsed beyond the Rhine might find a good base on Befort or Besancon, on Mezieres or Sedan, as the Russian army after the evacuation of Moscow left the base on the north and east and established itself upon the line of the Oka and the southern provinces. These lateral bases perpendicular to the front of defense are often decisive in preventing the enemy from penetrating to the heart of the country, or at least in rendering it impossible for him to maintain himself there. A base upon a broad and rapid river, both banks being held by strong works, would be as favorable as could be desired.

The more extended the base, the more difficulty will there be in covering it; but it will also be more difficult to cut the army off from it. A state whose capital is too near the frontier cannot have so favorable a base in a defensive war as one whose capital is more retired.

A base, to be perfect, should have two or three fortified points of sufficient capacity for the establishment of depots of supply. There should be a *tete de pont* upon each of its unfordable streams.

All are now agreed upon these principles; but upon other points opinions have varied. Some have asserted that a perfect base is one parallel to that of the enemy. My opinion is that bases perpendicular to those of the enemy are more advantageous, particularly such as have two sides almost perpendicular to each other and forming a re-entrant angle, thus affording a double base if required, and which, by giving the control of two sides of the strategic field, assure two lines of retreat widely apart, and facilitate any change of the line of operations which an unforeseen turn of affairs may necessitate.

The quotations which follow are from my treatise on Great Military Operations:–

"The general configuration of the theater of war may also have a great influence upon the direction of the lines of operations, and, consequently, upon the direction of the bases.

Figure 1

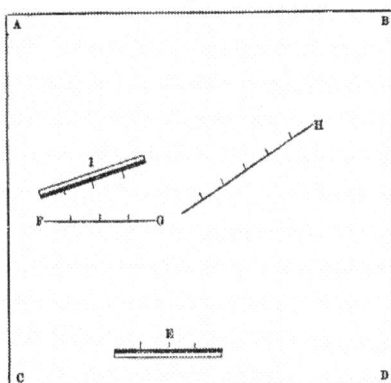

"If every theater of war forms a figure presenting four faces more or less regular, one of the armies, at the opening of the campaign, may hold one of these faces,—perhaps two,—while the enemy occupies the other, the fourth being closed by insurmountable obstacles. The different ways of occupying this theater will lead to widely different combinations. To illustrate, we will cite the theater of the French armies in Westphalia from 1757 to 1762, and that of Napoleon in 1806, both of which are represented in Fig. 1, p. 79. In the first case, the side A B was the North Sea, B D the line of the Weser and the base of Duke Ferdinand, C D the line of the Main and the base of the French army, A C the line of the Rhine, also guarded by French troops. The French held two faces, the North Sea being the third; and hence it was only necessary for them, by maneuvers, to gain the side B D to be masters of the four faces, including the base and the communications of the enemy. The French army, starting from its base C D and gaining the front of operations F G H, could cut off the allied army I from its base B D; the latter would be thrown upon the angle A, formed by the lines of the Rhine, the Ems, and the sea, while the army E could communicate with its bases on the Main and Rhine.

"The movement of Napoleon in 1806 on the Saale was similar. He occupied at Jena and Naumburg the line F G H, then marched by Halle and Dessau to force the Prussian army I upon the sea, represented by the side A B. The result is well known.

"The art, then, of selecting lines of operations is to give them such directions as to seize the communications of the enemy without losing one's own. The line F G H, by its extended position, and the bend on the flank of the enemy, always protects the communications with the base C D; and this is exactly the maneuvers of Marengo, Ulm, and Jena.

"When the theater of war does not border upon the sea, it is always bounded by a powerful neutral state, which guards its frontiers and closes one side of the square. This may not be an obstacle insurmountable like the sea; but generally it may be considered as an obstacle upon which it would be dangerous to retreat after a defeat: hence it would be an advantage to force the enemy upon it. The soil of a power which can bring into the field one hundred and fifty or two hundred thousand troops cannot be violated with impunity; and if a defeated army made the attempt, it would be none the less cut off from its base. If the boundary of the theater of war should be the territory of a weak state, it would be absorbed in this theater, and the square would be enlarged till it reached the frontiers of a powerful state, or the sea. The outline of the frontiers may modify the shape of the quadrilateral so as to make it approach the figure of a parallelogram or trapezoid, as in Figure 2. In either case, the advantage of the army which has control of two faces of the figure, and possesses the power of establishing upon them a double base, will be still more decided, since it will be able more easily to cut the enemy off from the shortened side,— as was the case with the Prussian army in 1806, with the side B D J of the parallelogram formed by the lines of the Rhine, the Oder, the North Sea, and the mountainous frontier of Franconia."

Figure 2

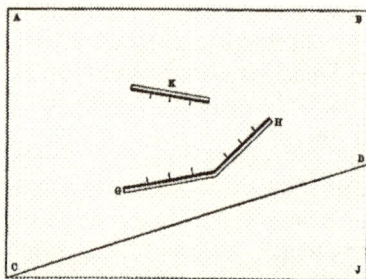

The selection of Bohemia as a base in 1813 goes to prove the truth of my opinion; for it was the perpendicularity of this base to that of the French army which enabled the allies to neutralize the immense advantages which the line of the Elbe would otherwise have afforded Napoleon, and turned the advantages of the campaign in their favor. Likewise, in 1812, by establishing their base perpendicularly upon the Oka and Kalouga, the Russians were able to execute their flank march upon Wiazma and Krasnoi.

If any thing further be required to establish these truths, it will only be necessary to consider that, if the base be perpendicular to that of the enemy, the front of operations will be parallel to his line of operations, and that hence it will be easy to attack his communications and line of retreat.

It has been stated that perpendicular bases are particularly favorable in the case of a double frontier, as in the last figures. Critics may object to this that it does not agree with what is elsewhere said in favor of frontiers which are salient toward the enemy, and against double lines of operations with equality of force. (Art. XXI.) The objection is not well founded; for the greatest advantage of a perpendicular base consists in the fact that it forms such a salient, which takes in reverse a portion of the theater of operations. On the other hand, a base with two faces by no means requires that both should be occupied in force: on the contrary, upon one of them it will be sufficient to have some fortified points garrisoned by small bodies, while the great bulk of the force rests upon the other face,—as was done in the campaigns of 1800 and 1806. The angle of nearly ninety degrees formed by the portion of the Rhine from Constance to Basel, and thence to Kehl, gave General Moreau one base parallel and another perpendicular to that of his antagonist. He threw two divisions by his left toward Kehl on the first base, to attract the attention of the enemy to that point, while he moved with nine divisions upon the extremity of the perpendicular face toward Schaffhausen, which carried him in a few days to the gates of Augsburg, the two detached divisions having already rejoined him.

In 1806, Napoleon had also the double base of the Rhine and Main, forming almost a right re-entrant angle. He left Mortier upon the first and parallel one, while with the mass of his forces he gained the extremity of the perpendicular base, and thus intercepted the Prussians at Gera and Naumburg by reaching their line of retreat.

If so many imposing facts prove that bases with two faces, one of them being almost perpendicular to that of the enemy, are the best, it is well to recollect that, in default of such a base, its advantages may be partially supplied by a change of strategic front, as will be seen in Article XX.

Another very important point in reference to the proper direction of bases relates to those established on the sea-coast. These bases may be favorable in some circumstances, but are equally unfavorable in others, as may be readily seen from what precedes. The danger which must always exist of an army being driven to the sea seems so clear, in the ease of the establishment of the base upon it, (which bases can only be favorable to naval powers,) that it is astonishing to hear in our day praises of such a base. Wellington, coming with a fleet to the relief of Spain and Portugal, could not have secured a better base than that of Lisbon, or rather of the peninsula of Torres-Vedras, which covers all the avenues to that capital on the land side. The sea and the Tagus not only protected both flanks, but secured the safety of his only possible line of retreat, which was upon the fleet.

Blinded by the advantages which the intrenched camp of Torres-Vedras secured for the English, and not tracing effects to their real causes, many generals in other respects wise contend that no bases are good except such as rest on the sea and thus afford the army facilities of supply and refuge with both flanks secured. Fascinated by similar notions, Colonel Carion-Nizas asserted that in 1813 Napoleon ought to have posted half of his army in Bohemia and thrown one hundred and fifty thousand men on the mouths of the Elbe toward Hamburg; forgetting that the first precept for a continental army is to establish its base upon the front farthest *from* the sea, so as to secure the benefit of all its elements of strength, from which it might find itself cut off if the base were established upon the coast.

An insular and naval power acting on the continent would pursue a diametrically opposite course, but resulting from the same principle, viz.: *to establish the base upon those points where it can be sustained by all the resources of the country, and at the same time insure a safe retreat.*

A state powerful both on land and sea, whose squadrons control the sea adjacent to the theater of operations, might well base an army of forty or fifty thousand men upon the coast, as its retreat by sea and its supplies could be well assured; but to establish a continental army of one hundred and fifty thousand men upon such a base, when opposed by a disciplined and nearly equal force, would be an act of madness.

However, as every maxim has its exceptions, there is a case in which it may be admissible to base a continental army upon the sea: it is, when your adversary is not formidable upon land, and when you, being master of the sea, can supply the army with more facility than in the interior. We rarely see these conditions fulfilled: it was so, however, during the Turkish war of 1828 and 1829. The whole attention of the Russians was given to Varna and Bourghas, while Shumla was merely observed; a plan which they could not have pursued in the presence of a European army (even with the control of the sea) without great danger of ruin.

Despite all that has been said by triflers who pretend to decide upon the fate of empires, this war was, in the main, well conducted. The army covered itself by obtaining the fortresses of Brailoff, Varna, and Silistria, and afterward by preparing a depot at Sizeboli. As soon as its base was well established it moved upon Adrianople, which previously would have been madness. Had the season been a couple of months longer, or had the army not come so great a distance in 1828, the war would have terminated with the first campaign.

Besides permanent bases, which are usually established upon our own frontiers, or in the territory of a faithful ally, there are eventual or temporary bases, which result from the operations in the enemy's country; but, as these are rather temporary points of support, they will, to avoid confusion, be discussed in Article XXIII.

ARTICLE XIX

STRATEGIC LINES AND POINTS, DECISIVE POINTS OF THE THEATER OF WAR, AND OBJECTIVE POINTS OF OPERATIONS

Strategic lines and points are of different kinds. Some receive this title simply from their position, which gives them all their importance: these are permanent geographical strategic points. Others have a value from the relations they bear to the positions of the masses of the hostile troops and to the enterprises likely to be directed against them: such are strategic points of maneuver, and are eventual. Finally, there are points which have only a secondary importance, and others whose importance is constant and immense: the latter are called DECISIVE strategic points.

Every point of the theater of war which is of military importance, whether from its position as a center of communication, or from the presence of military establishments or fortifications, is a geographical strategic point.

A distinguished general affirms that such a point would not necessarily be a strategic point, unless situated favorably for a contemplated operation. I think differently; for a strategic point is such essentially and by nature, and, no matter how far distant it may be from the scene of the first enterprises, it may be included in the field by some unforeseen turn of events, and thus acquire its full importance. It would, then, be more accurate to state that all strategic points are not necessarily decisive points.

Lines are strategic either from their geographical position or from their relation to temporary maneuvers. The first class may be subdivided as follows,—viz.: geographic lines which by their permanent importance belong to the decisive points[1] of the theater of war, and those which have value merely because they connect two strategic points.

To prevent confusion, we will elsewhere treat of strategic lines in their relations to maneuvers,—confining ourselves here to what relates to the *decisive and objective points* of the zone of operations upon which enterprises occur.

Although these are most intimately connected, since every objective point ought necessarily to be one of the decisive points of the theater of war, there is nevertheless a distinction between them; for all decisive points cannot be at the same time the objective of operations. We will, then, define the first, in order to be more easily guided in our selection of the second.

---

[1]   I may be reproached with inaccuracy of expression,—since a line cannot be a *point*, and yet I apply to lines the name of decisive or objective points. It seems almost useless to remark that *objective* points are not geometric points, but that the name is a form of expression used to designate the object which an army desires to attain.

I think the name of *decisive strategic point* should be given to all those which are capable of exercising a marked influence either upon the result of the campaign or upon a single enterprise. All points whose geographical position and whose natural or artificial advantages favor the attack or defense of a front of operations or of a line of defense are included in this number; and large, well-located fortresses occupy in importance the first rank among them.

The decisive points of a theater of war are of several kinds. The first are the geographic points and lines whose importance is permanent and a consequence of the configuration of the country. For example, take the case of the French in Belgium: whoever is master of the line of the Meuse will have the greatest advantages in taking possession of the country; for his adversary, being outflanked and inclosed between the Meuse and the North Sea, will be exposed to the danger of total ruin if he give battle parallel to that sea.[1] Similarly, the valley of the Danube presents a series of important points which have caused it to be looked upon as the key of Southern Germany.

Those points the possession of which would give the control of the junction of several valleys and of the center of the chief lines of communication in a country are also *decisive geographic points*. For instance, Lyons is an important strategic point, because it controls the valleys of the Rhone and Saone, and is at the center of communications between France and Italy and between the South and East; but it would not be a *decisive* point unless well fortified or possessing an extended camp with *tetes de pont*. Leipsic is most certainly a strategic point, inasmuch as it is at the junction of all the communications of Northern Germany. Were it fortified and did it occupy both banks of the river, it would be almost the key of the country,—if a country has a key, or if this expression means more than a decisive point.

All capitals are strategic points, for the double reason that they are not only centers of communications, but also the seats of power and government.

In mountainous countries there are defiles which are the only routes of exit practicable for an army; and these may be decisive in reference to any enterprise in this country. It is well known how great was the importance of the defile of Bard, protected by a single small fort, in 1800.

The second kind of decisive points are accidental points of maneuver, which result from the positions of the troops on both sides.

When Mack was at Ulm, in 1805, awaiting the approach of the Russian army through Moravia, the decisive point in an attack upon him was Donauwerth or the Lower Lech; for if his adversaries gained it before him he was cut off from his line of retreat, and also from the army intended to support him. On the contrary, Kray, who, in 1800, was in the same position, expected no aid from Bohemia, but rather from the Tyrol and from the army of Melas in Italy: hence the decisive point of attack upon him was not Donauwerth, but on the opposite side, by Schaffhausen, since this would take in reverse his front of

---

[1]   This only applies to continental armies, and not to the English, who, having their base on Antwerp or Ostend, would have nothing to fear from an occupation of the line of the Meuse.

operations, expose his line of retreat, cut him off from his supporting army as well as from his base, and force him upon the Main. In the same campaign the first objective point of Napoleon was to fall upon the right of Melas by the Saint-Bernard, and to seize his line of communications: hence Saint-Bernard, Ivrea, and Piacenza were decisive points only by reason of the march of Melas upon Nice.

It may be laid down as a general principle that the decisive points of maneuver are on that flank of the enemy upon which, if his opponent operates, he can more easily cut him off from his base and supporting forces without being exposed to the same danger. The flank opposite to the sea is always to be preferred, because it gives an opportunity of forcing the enemy upon the sea. The only exception to this is in the case of an insular and inferior army, where the attempt, although dangerous, might be made to cut it off from the fleet.

If the enemy's forces are in detachments, or are too much extended, the decisive point is his center; for by piercing that, his forces will be more divided, their weakness increased, and the fractions may be crushed separately.

The decisive point of a battle-field will be determined by,—

1. The features of the ground.
2. The relation of the local features to the ultimate strategic aim.
3. The positions occupied by the respective forces.

These considerations will be discussed in the chapter on battles.

OBJECTIVE POINTS

There are two classes of objective points,—objective *points of maneuver*, and *geographical objective points*. A geographical objective point may be an important fortress, the line of a river, a front of operations which affords good lines of defense or good points of support for ulterior enterprises. *Objective points of maneuver*, in contradistinction to *geographical objectives*, derive their importance from, and their positions depend upon, the situation of the hostile masses.

In strategy, the object of the campaign determines the objective point. If this aim be offensive, the point will be the possession of the hostile capital, or that of a province whose loss would compel the enemy to make peace. In a war of invasion the capital is, ordinarily, the objective point. However, the geographical position of the capital, the political relations of the belligerents with their neighbors, and their respective resources, are considerations foreign in themselves to the art of fighting battles, but intimately connected with plans of operations, and may decide whether an army should attempt or not to occupy the hostile capital. If it be concluded not to seize the capital, the objective point might be a part of the front of operations or line of defense where an important fort is situated, the possession of which would render safe the occupation of the neighboring territory. For instance, if France were to invade Italy in a war against Austria, the first objective point would be the line of the Ticino and Po; the second, Mantua and the line of the Adige. In the defensive, the objective point, instead of being that which it is desirable to gain possession of, is that which is to be defended. The capital, being considered the seat of power, becomes the principal objective point of the defense; but there may be other points, as the defense of a first line and of the first base of operations. Thus, for a French army reduced to the defensive behind the Rhine, the first objective would be to prevent the passage of the river; it would endeavor to relieve the forts in Alsace if the enemy succeeded in effecting a passage of the river and in besieging them: the second objective would be to cover the first base of operations upon the Meuse or Moselle,—which might be attained by a lateral defense as well as one in front.

As to the objective points of *maneuvers*,—that is, those which relate particularly to the destruction or decomposition of the hostile forces,—their importance may be seen by what has already been said. The greatest talent of a general, and the surest hope of success, lie in some degree in the good choice of these points. This was the most conspicuous merit of Napoleon. Rejecting old systems, which were satisfied by the capture of one or two points or with the occupation of an adjoining province, he was convinced that the best means of accomplishing great results was to dislodge and destroy the hostile army,— since states and provinces fall of themselves when there is no organized force to protect them. To detect at a glance the relative advantages presented by the different zones of operations, to concentrate the mass of the forces upon that one which gave the best promise of success, to be indefatigable in ascertaining the approximate position of the enemy, to fall with the rapidity of lightning upon his center if his front was too much extended, or upon that flank by which he could more readily seize his communications, to outflank him, to cut his line, to pursue him to the last, to disperse and destroy his forces,—such was the system followed by Napoleon in his first campaigns. These campaigns proved this system to be one of the very best.

When these maneuvers were applied, in later years, to the long distances and the inhospitable regions of Russia, they were not so successful as in Germany: however, it must be remembered that, if this kind of war is not suitable to all capacities, regions, or circumstances, its chances of success are still very great, and it is based upon principle. Napoleon abused the system; but this does not disprove its real advantages when a proper limit is assigned to its enterprises and they are made in harmony with the respective conditions of the armies and of the adjoining states.

The maxims to be given on these important strategic operations are almost entirely included in what has been said upon decisive points, and in what will be stated in Article XXI. in discussing the choice of lines of operations.

As to the choice of objective points, every thing will generally depend upon the aim of the war and the character which political or other circumstances may give it, and, finally, upon the military facilities of the two parties.

In cases where there are powerful reasons for avoiding all risk, it may be prudent to aim only at the acquisition of partial advantages,—such as the capture of a few towns or the possession of adjacent territory. In other cases, where a party has the means of achieving a great success by incurring great dangers, he may attempt the destruction of the hostile army, as did Napoleon.

The maneuvers of Ulm and Jena cannot be recommended to an army whose only object is the siege of Antwerp. For very different reasons, they could not be recommended to the French army beyond the Niemen, five hundred leagues from its frontiers, because there would be much more to be lost by failure than a general could reasonably hope to gain by success.

There is another class of decisive points to be mentioned, which are determined more from political than from strategic considerations: they play a great part in most coalitions, and influence the operations and plans of cabinets. They may be called *political objective points.*

Indeed, besides the intimate connection between statesmanship and war in its preliminaries, in most campaigns some military enterprises are undertaken to carry out a political end, sometimes quite important, but often very irrational. They frequently lead to the commission of great errors in strategy. We cite two examples. First, the expedition of the Duke of York to Dunkirk, suggested by old commercial views, gave to the operations of the allies a divergent direction, which caused their failure: hence this objective point was bad in a military view. The expedition of the same prince to Holland in 1799—likewise due to the views of the English cabinet, sustained by the intentions of Austria on Belgium—was not less fatal; for it led to the march of the Archduke Charles from Zurich upon Manheim,—a step quite contrary to the interests of the allied armies at the time it was undertaken. These illustrations prove that political objective points should be subordinate to strategy, at least until after a great success has been attained.

This subject is so extensive and so complicated that it would be absurd to attempt to reduce it to a few rules. The only one which can be given has just been alluded to, and is, that either the political objective points should be selected according to the principles of strategy, or their consideration should be postponed till after the decisive events of the campaign. Applying this rule to the examples just given, it will be seen that it was at Cambray or in the heart of France that Dunkirk should have been conquered in 1793 and Holland delivered in 1799; in other words, by uniting all the strength of the allies for great attempts on the decisive points of the frontiers. Expeditions of this kind are generally included in grand diversions,—to be treated of in a separate article.

ARTICLE XX

FRONTS OF OPERATIONS, STRATEGIC FRONTS, LINES OF DEFENSE, AND STRATEGIC POSITIONS

There are some parts of the military science that so closely resemble each other, and are so intimately allied, that they are frequently confounded, although they are decidedly distinct. Such are *fronts of operations, strategic fronts, lines of defense,* and *strategic positions.* It is proposed in this article to show the distinction between them and to expose their relations to each other.

FRONTS OF OPERATIONS AND STRATEGIC FRONTS

When the masses of an army are posted in a zone of operations, they generally occupy strategic positions. The extent of the front occupied toward the enemy is called the *strategic front.* The portion of the theater of war from which an enemy can probably reach this front in two or three marches is called the *front of operations.*

The resemblance between these two fronts has caused many military men to confound them, sometimes under one name and sometimes under the other.

Rigorously speaking, however, the strategic front designates that formed by the actual positions occupied by the masses of the army, while the other embraces the space separating the two armies, and extends one or two marches beyond each extremity of the strategic front, and includes the ground upon which the armies will probably come in collision.

When the operations of a campaign are on the eve of commencing, one of the armies will decide to await the attack of the other, and will undertake to prepare a line of defense, which may be either that of the strategic front or more to the rear. Hence the strategic front and line of defense may coincide, as was the case in 1795 and 1796 upon the Rhine, which was then a line of defense for both Austrians and French, and at the same time their strategic front and front of operations. This occasional coincidence of these lines doubtless leads persons to confound them, while they are really very different. An army has not necessarily a line of defense, as, for example, when it invades: when its masses are concentrated in a single position, it has no strategic front, but it is never without a front of operations.

The two following examples will illustrate the difference between the different terms.

At the resumption of hostilities in 1813, Napoleon's front of operations extended at first from Hamburg to Wittenberg; thence it ran along the line of the allies toward Glogau and Breslau, (his right being at Loewenberg,) and followed along the frontier of Bohemia to Dresden. His forces were stationed on this grand front in four masses, whose strategic positions were interior and central and presented three different faces. Subsequently, he retired behind the Elbe. His real line of defense then extended only from Wittenberg to Dresden, with a bend to the rear toward Marienberg, for Hamburg and Magdeburg were beyond the strategic field, and it would have been fatal for him to have extended his operations to these points.

The other example is his position about Mantua in 1796. His front of operations here really extended from the mountains of Bergamo to the Adriatic Sea, while his real line of defense was upon the Adige, between Lake Garda and Legnago: afterward it was upon the Mincio, between Peschiera and Mantua, while his strategic front varied according to his positions.

The front of operations being the space which separates the two armies, and upon which they may fight, is ordinarily parallel to the base of operations. The strategic front will have the same direction, and ought to be perpendicular to the principal line of operations, and to extend far enough on either flank to cover this line well. However, this direction may vary, either on account of projects that are formed, or on account of the attacks of the enemy; and it quite frequently happens that it is necessary to have a front perpendicular to the base and parallel to the original line of operations. Such a change of strategic front is one of the most important of all grand maneuvers, for by this means the control of two faces of the strategic field may be obtained, thus giving the army a position almost as favorable as if it possessed a base with two faces. (See Art. XVIII.)

The strategic front of Napoleon in his march on Eylau illustrates these points. His pivots of operations were at Warsaw and Thorn, which made the Vistula a temporary base: the front became parallel to the Narew, from whence he set out, supported by Sierock, Pultusk, and Ostrolenka, to maneuver by his right and throw the Russians on Elbing and the Baltic. In such cases, if a point of support in the new direction can be obtained, the strategic front gives the advantages referred to above. It ought to be borne in mind in such maneuvers that the army should always be sure of regaining its temporary base if necessary; in other words, that this base should be prolonged behind the strategic front, and should be covered by it. Napoleon, marching from the Narew by Allenstein upon Eylau, had behind his left Thorn, and farther from the front of the army the *tete de pont* of Praga and Warsaw; so that his communications were safe, while Benningsen, forced to face him and to make his line parallel to the Baltic, might be cut off from his base, and be thrown back upon the mouths of the Vistula. Napoleon executed another very remarkable change of strategic front in his march from Gera upon Jena and Naumburg in 1806. Moreau made another in moving by his right upon Augsburg and Dillingen, fronting the Danube and France, and thereby forcing Kray to evacuate the intrenched camp at Ulm.

The change of the strategic front to a position perpendicular to the base may be a temporary movement for an operation of a few days' duration, or it may be for an indefinite time, in order to profit by important advantages afforded by certain localities, to strike decisive blows, or to procure for the army a good line of defense and good pivots of operations, which would be almost equivalent to a real base.

It often happens that an army is compelled to have a double strategic front, either by the features of the theater of war, or because every line of offensive operations requires protection on its flanks. As an example of the first, the frontiers of Turkey and Spain may be cited. In order to cross the Balkan or the Ebro, an army would be obliged to present a double front,—in the first case, to face the valley of the Danube; in the second, to confront forces coming from Saragossa or Leon.

All extensive countries necessitate, to a greater or less degree, the same precaution. A French army in the valley of the Danube will require a double front as soon as the Austrians have thrown sufficient troops into the Tyrol or Bohemia to give rise to any anxiety. Those countries which present a narrow frontier to the enemy are the only exception, since the troops left on the frontier to harass the flanks of the enemy could themselves be cut off and captured. This necessity of double strategic fronts is one of the most serious inconveniences of an offensive war, since it requires large detachments, which are always dangerous. (See Article XXXVI.)

Of course, all that precedes relates to regular warfare. In a national or intestine war the whole country is the scene of hostilities. Nevertheless, each large fraction of an army having a defined aim would have its own strategic front determined by the features of the country and the positions occupied by the large bodies of the enemy. Thus, Suchet in Catalonia and Massena in Portugal each had a strategic front, while the front of some other corps of the army was not clearly defined.

LINES OF DEFENSE

Lines of defense are classified as strategical and tactical. Strategical lines of defense are subdivided into two classes: 1. Permanent lines of defense, which are a part of the defensive system of a state, such as the line of a fortified frontier; 2. Eventual lines of defense, which relate only to the temporary position of an army.

The frontier is a permanent line of defense when it presents a well-connected system of obstacles, natural and artificial, such as ranges of mountains, broad rivers, and fortresses. Thus, the range of the Alps between France and Piedmont is a line of defense, since the practicable passes are guarded by forts which would prove great obstacles in the way of an army, and since the outlets of the gorges in the valleys of Piedmont are protected by large fortresses. The Rhine, the Oder, and the Elbe may also be considered as permanent lines of defense, on account of the important forts found upon them.

Every river of any considerable width, every range of mountains, and every defile, having their weak points covered by temporary fortifications, may be regarded as *eventual lines of defense*, both strategic and tactical, since they may arrest for some time the progress of the enemy, or may compel him to deviate to the right or left in search of a weaker point,—in which case the advantage is evidently strategic. If the enemy attack in front, the lines present an evident tactical advantage, since it is always more difficult to drive an army from its position behind a river, or from a point naturally and artificially strong, than to attack it on an open plain. On the other hand, this advantage must not be considered unqualified, lest we should fall into the system of positions which has been the ruin of so many armies; for, whatever may be the facilities of a position for defense, it is quite certain that the party which remains in it passive and receiving all the attacks of his adversary will finally yield.[I] In addition to this, since a position naturally very strong[II] is difficult of access it will be as difficult of egress, the enemy may be able with an inferior force to confine the army by guarding all the outlets. This happened to the Saxons in the camp of Pirna, and to Wurmser in Mantua.

STRATEGIC POSITIONS

There is a disposition of armies to which the name of strategic position may be applied, to distinguish from tactical positions or positions for battle.

Strategic positions are those taken for some time and which are intended to cover a much greater portion of the front of operations than would be covered in an actual battle. All positions behind a river or upon a line of defense, the divisions of the army being separated by considerable distances, are of this class, such as those of Napoleon at Rivoli, Verona, and Legnago to overlook the Adige. His positions in 1813 in Saxony and Silesia in advance of his line of defense were strategic. The positions of the Anglo-Prussian armies on the frontier of Belgium before the battle of Ligny, (1814,) and that of Massena on the Limmat and Aar in 1799, were also strategic. Even winter quarters, when compact and in face of the enemy and not protected by an armistice, are strategic positions,—for instance, Napoleon on the Passarge in 1807. The daily positions taken up by an army beyond the reach of the enemy, which are sometimes spread out either to deceive him or to facilitate movements, are of this class.

This class also includes positions occupied by an army to cover several points and positions held by the masses of an army for the purposes of observation. The different positions taken up on a line of defense, the positions of detachments on a double front of operations, the position of a detachment covering a siege,

---

[I]   This does not refer to intrenched camps, which make a great difference. They are treated of in Article XXVII.
[II]   It is a question here of positions of camps, and not of positions for battle. The latter will be treated of in the chapter devoted to Grand Tactics, (Article XXX.)

the main army in the meanwhile operating on another point, are all strategic. Indeed, all large detachments or fractions of an army may be considered as occupying strategic positions.

The maxims to be given on the preceding points are few, since fronts, lines of defense, and strategic positions generally depend upon a multitude of circumstances giving rise to infinite variety.

In every case, the first general rule is that the communications with the different points of the line of operations be thoroughly assured.

In the defense it is desirable that the strategic fronts and lines of defense should present both upon the flanks and front formidable natural or artificial obstacles to serve as points of support. The points of support on the strategic front are called *pivots of operations*, and are practical temporary bases, but quite different from pivots of maneuver. For example, in 1796 Verona was an excellent pivot of operations for all Napoleon's enterprises about Mantua for eight months. In 1813 Dresden was his pivot.

Pivots of maneuver are detachments of troops left to guard points which it is essential to hold, while the bulk of the army proceeds to the fulfillment of some important end; and when this is accomplished the pivot of maneuver ceases to exist. Thus, Ney's corps was the pivot of Napoleon's maneuver by Donauwerth and Augsburg to cut Mack from his line of retreat. A pivot of operations, on the contrary, is a material point of both strategical and tactical importance, serves as a point of support and endures throughout a campaign.

The most desirable quality of a line of defense is that it should be as short as possible, in order to be covered with facility by the army if it is compelled to take the defensive. It is also important that the extent of the strategic front should not be so great as to prevent the prompt concentration of the fractions of the army upon an advantageous point.

The same does not altogether apply to the front of operations; for if it be too contracted it would be difficult for an army on the offensive to make strategic maneuvers calculated to produce great results, since a short front could be easily covered by the defensive army. Neither should the front of operations be too extended. Such a front is unsuitable for offensive operations, as it would give the enemy, if not a good line of defense, at least ample space to escape from the results of a strategic maneuver even if well planned. Thus, the beautiful operations of Marengo, Ulm, and Jena could not have produced the same results upon a theater of the magnitude of that of the Russian War in 1812, since the enemy, even if cut off from his line of retreat, could have found another by adopting a new zone of operations.

The essential conditions for every strategic position are that it should be more compact than the forces opposed, that all fractions of the army should have sure and easy means of concentrating, free from the intervention of the enemy. Thus, for forces nearly equal, all central or interior positions would be preferable to exterior ones, since the front in the latter case would necessarily be more extended and would lead to a dangerous division of force. Great mobility and activity on the part of the troops occupying these positions will be a strong element of security or of superiority over the enemy, since it renders possible rapid concentration at different and successive points of the front.

An army should never long occupy any strategic point without making selection of one or two tactical positions, for the purpose of there concentrating all the disposable force, and giving battle to the enemy when he shall have unveiled his designs. In this manner Napoleon prepared the fields of Rivoli and Austerlitz, Wellington that of Waterloo, and the Archduke Charles that of Wagram.

When an army either camps or goes into quarters, the general should be careful that the front be not too extended. A disposition which might be called the strategic square seems best, presenting three nearly-equal faces, so that the distance to be passed over would be about equal for all the divisions in concentrating upon the common center to receive an attack.

Every strategic line of defense should always possess a tactical point upon which to rally for defense should the enemy cross the strategic front. For instance, an army guarding a bank of a river, not being able to occupy in force the whole line, ought always to have a position in rear of the center selected, upon which to collect all his divisions, so as to oppose them united to the enemy when he has succeeded in effecting a passage.

For an army entering a country with the purpose either of subjugation or of temporary occupation, it would always be prudent, however brilliant may have been its earlier successes, to prepare a line of defense as a refuge in case of reverse. This remark is made to complete the subject: the lines themselves are intimately connected with temporary bases, and will be discussed in a future article, (XXIII.)

ARTICLE XXI

ZONES AND LINES OF OPERATIONS

A zone of operations is a certain fraction of the whole theater of war, which may be traversed by an army in the attainment of its object, whether it act singly or in concert with other and secondary armies. For example, in the plan of campaign of 1796, Italy was the zone of the right, Bavaria that of the center, Franconia that of the left army.

A zone of operations may sometimes present but a single *line of operations*, either on account of the configuration of the country, or of the small number of practicable routes for an army found therein. Generally, however, a zone presents several *lines of operations*, depending partly upon the plans of the campaign, partly upon the number of great routes of communication existing in the theater of operations.

It is not to be understood from this that every road is of itself a *line of operations*,—though doubtless it may happen that any good road in a certain turn of affairs may become for the time-being such a line; but as long as it is only traversed by detachments, and lies beyond the sphere of the principal enterprises, it cannot truly be called the real line of operations. Moreover, the existence of several routes leading to the same front of operations, and separated by one or two marches, would not constitute so many lines of operations, but, being the communications of the different divisions of the same army, the whole space bounded by them would constitute but a single line.

The term *zone of operations* is applied to a large fraction of the general theater of war; the term *lines of operations* will designate the part of this fraction embraced by the enterprises of the army. Whether it follow a single or several routes, the term *strategic lines* will apply to those important lines which connect the decisive points of the theater of operations either with each other or with the front of operations; and, for the same reason, we give this name to those lines which the army would follow to reach one of these decisive points, or to accomplish an important maneuver which requires a temporary deviation from the principal line of operations. *Lines of communications* designate the practicable routes between the different portions of the army occupying different positions throughout the zone of operations.

For example, in 1813, after the accession of Austria to the Grand Coalition, three allied armies were to invade Saxony, one Bavaria, and another Italy: so that Saxony, or rather the country between Dresden, Magdeburg, and Breslau, formed the zone of operations of the mass of the forces. This zone had three *lines of operations* leading to Leipsic as an objective: the first was the line of the army of Bohemia, leading from the mountains of Erzgebirge by Dresden and Chemnitz upon Leipsic; the second was the line of the army of Silesia, going from Breslau by Dresden or by Wittenberg upon Leipsic; the third was that of Bernadotte from Berlin by Dessau to the same objective point. Each of these armies marched upon two or more adjacent parallel routes, but it could not be said that there were as many lines of operations as roads. The principal line of operations is that followed by the bulk of the army, and upon which depots of provisions, munitions, and other supplies are echeloned, and over which, if compelled, it would retreat.

If the choice of a zone of operations involves no extensive combinations, since there can never be more than two or three zones on each theater, and the advantages generally result from the localities, it is somewhat different with lines of operations, as they are divided into different classes, according to their relations to the different positions of the enemy, to the communications upon the strategic field, and to the enterprises projected by the commander.

Simple lines of operations are those of an army acting from a frontier when it is not subdivided into large independent bodies.

Double lines of operations are those of two independent armies proceeding from the same frontier, or those of two nearly equal armies which are commanded by the same general but are widely separated in distance and for long intervals of time.[I]

Interior lines of operations are those adopted by one or two armies to oppose several hostile bodies, and having such a direction that the general can concentrate the masses and maneuver with his whole force in a shorter period of time than it would require for the enemy to oppose to them a greater force.[II] *Exterior lines* lead to the opposite result, and are those formed by an army which operates at the same time on both flanks of the enemy, or against several of his masses.

Concentric lines of operations are those which depart from widely-separated points and meet at the same point, either in advance of or behind the base.

Divergent lines are those by which an army would leave a given point to move upon several distinct points. These lines, of course, necessitate a subdivision of the army.

There are also *deep lines*, which are simply *long lines*.

The term *maneuver-lines* I apply to momentary strategic lines, often adopted for a single temporary maneuver, and which are by no means to be confounded with the real *lines of operations*.

Secondary lines are those of two armies acting so as to afford each other mutual support,—as, in 1796, the army of the Sambre and Meuse was secondary to the army of the Rhine, and, in 1812, the army of Bagration was secondary to that of Barclay.

Accidental lines are those brought about by events which change the original plan and give a new direction to operations. These are of the highest importance. The proper occasions for their use are fully recognized only by a great and active mind.

There may be, in addition, *provisional* and *definitive lines of operations*. The first designate the line adopted by an army in a preliminary, decisive enterprise, after which it is at liberty to select a more advantageous or direct line. They seem to belong as much to the class of temporary or eventual strategic lines as to the class of lines of operations.

I   This definition has been criticized; and, as it has given rise to misapprehension, it becomes necessary to explain it.
In the first place, it must be borne in mind that it is a question of *maneuver-lines*, (that is, of strategic combinations,) and not of great routes. It must also be admitted that an army marching upon two or three routes, near enough to each other to admit of the concentration of the different masses within forty-eight hours, would not have two or three lines of operations. When Moreau and Jourdan entered Germany with two armies of 70,000 men each, being independent of each other, there was a double line of operations; but a French army of which only a detachment starts from the Lower Rhine to march on the Main, while the five or six other corps set out from the Upper Rhine to march on Ulm, would not have a double line of operations in the sense in which I use the term to designate a maneuver. Napoleon, when he concentrated seven corps and set them in motion by Bamberg to march on Gera, while Mortier with a single corps marched on Cassel to occupy Hesse and flank the principal enterprise, had but a single general line of operations, with an accessory detachment. The territorial line was composed of two arms or radii, but the operation was not double.
II   Some German writers have said that I confound central positions with the line of operations,—in which assertion they are mistaken. An army may occupy a central position in the presence of two masses of the enemy, and not have interior lines of operations: these are two very different things. Others have thought that I would have done better to use the term *radii of operations* to express the idea of double lines. The reasoning in this case is plausible if we conceive the theater of operations to be a circle; but, as every radius is, after all, a line, it is simply a dispute about words.

These definitions show how I differ from those authors who have preceded me. Lloyd and Bulow attribute to these lines no other importance than that arising from their relations to the depots of the army: the latter has even asserted that when an army is encamped near its depots it has no lines of operations.

The following example will disprove this paradox. Let us suppose two armies, the first on the Upper Rhine, the second in advance of Dusseldorf or any other point of this frontier, and that their large depots are immediately behind the river,—certainly the safest, nearest, and most advantageous position for them which could possibly be adopted. These armies will have an offensive or defensive object: hence they will certainly have lines of operations, arising from the different proposed enterprises.

1. Their defensive territorial line, starting from their positions, will extend to the second line which they are to cover, and they would both be cut off from this second line should the enemy establish himself in the interval which separates them from it. Even if Melas[1] had possessed a year's supplies in Alessandria, he would none the less have been cut off from his base of the Mincio as soon as the victorious enemy occupied the line of the Po.

2. Their line would be double, and the enemy's single if he concentrated his forces to defeat these armies successively; it would be a double exterior line, and the enemy's a double interior, if the latter divided his forces into two masses, giving them such directions as to enable him to concentrate all his forces before the two armies first referred to could unite.

Bulow would have been more nearly right had he asserted that an army on its own soil is less dependent on its primitive line of operations than when on foreign ground; for it finds in every direction points of support and some of the advantages which are sought for in the establishment of lines of operations; it may even lose its line of operations without incurring great danger; but that is no reason why it has no line of operations.

OBSERVATIONS UPON THE LINES OF OPERATIONS IN THE WARS OF THE FRENCH REVOLUTION

At the beginning of this terrible and ever-varying struggle, Prussia and Austria were the only avowed enemies of France, and Italy was included in the theater of war only for purposes of reciprocal observation, it being too remote for decisive enterprises in view of the end proposed. The real theater extended

---

[1]   This assertion has been disputed. I think it is correct; for Melas, confined between the Bormida, the Tanaro, and the Po, was unable to recruit for his army, barely able to maintain a communication by couriers with his base, and he certainly would have been obliged to cut his way out or to surrender in case he had not been reinforced.

from Huningue to Dunkirk, and comprised three zones of operations,—the first reaching along the Rhine from Huningue to Landau, and thence to the Moselle; the center consisting of the interval between the Meuse and Moselle; the third and left was the frontier from Givet to Dunkirk.

When France declared war, in April, 1792, her intention was to prevent a union of her enemies; and she had then one hundred thousand men in the zones just described, while Austria had but thirty-five thousand in Belgium. It is quite impossible to understand why the French did not conquer this country, when no effectual resistance could have been made. Four months intervened between the declaration of war and the concentration of the allied troops. Was it not probable that an invasion of Belgium would have prevented that of Champagne, and have given the King of Prussia a conception of the strength of France, and induced him not to sacrifice his armies for the secondary object of imposing upon France another form of government?

When the Prussians arrived at Coblentz, toward the end of July, the French were no longer able to invade. This *role* was reserved for the allies; and it is well known how they acquitted themselves.

The whole force of the French was now about one hundred and fifteen thousand men. It was scattered over a frontier of one hundred and forty leagues and divided into five corps d'armee, and could not make a good defense; for to paralyze them and prevent their concentration it was only necessary to attack the center. Political reasons were also in favor of this plan of attack: the end proposed was political, and could only be attained by rapid and vigorous measures. The line between the Moselle and Meuse, which was the center, was less fortified than the rest of the frontier, and, besides, gave the allies the advantage of the excellent fortress of Luxembourg as a base. They wisely adopted this plan of attack; but the execution was not equal to the conception.

The court of Vienna had the greatest interest in the war, for family reasons, as well as on account of the dangers to which a reverse might subject her provinces. For some reason, difficult to understand, Austria co-operated only to the extent of thirty battalions: forty-five thousand men remained as an army of observation in Brisgau, on the Rhine, and in Flanders. Where were the imposing armies she afterward displayed? and what more useful disposition could have been made of them than to protect the flanks of the invading army? This remarkable conduct on the part of Austria, which cost her so much, may account for the resolution of Prussia to retire at a later period, and quit the field, as she did, at the very moment when she should have entered it. During the campaign the Prussians did not exhibit the activity necessary for success. They spent eight days uselessly in camp at Kons. If they had anticipated Dumouriez at the Little Islands, or had even made a more serious effort to drive him from them, they would still have had all the advantage of a concentrated force against several scattered divisions, and could have prevented their junction and overthrown them separately. Frederick the Great would have justified the remark of Dumouriez at Grandpre,—that, if his antagonist had been the great king, he (Dumouriez) would already have been driven behind Chalons.

The Austrians in this campaign proved that they were still imbued with the false system of Daun and Lascy, of covering every point in order to guard every point.

The fact of having twenty thousand men in Brisgau while the Moselle and Sarre were uncovered, shows the fear they had of losing a village, and how their system led to large detachments, which are frequently the ruin of armies.

Forgetting that the surest hope of victory lies in presenting the strongest force, they thought it necessary to occupy the whole length of a frontier to prevent invasion,—which was exactly the means of rendering invasion upon every point feasible.

I will further observe that, in thin campaign, Dumouriez foolishly abandoned the pursuit of the allies in order to transfer the theater from the center to the extreme left of the general field. Moreover, he was unable to perceive the great results rendered possible by this movement, but attacked the army of the Duke of Saxe-Teschen in front, while by descending the Meuse to Namur he might have thrown it back upon the North Sea toward Meuport or Ostend, and have destroyed it entirely in a more successful battle than that of Jemmapes.

The campaign of 1793 affords a new instance of the effect of a faulty direction of operations. The Austrians were victorious, and recovered Belgium, because Dumouriez unskillfully extended his front of operations to the gates of Rotterdam. Thus far the conduct of the allies deserves praise: the desire of reconquering these rich provinces justified this enterprise, which, moreover, was judiciously directed against the extreme right of the long front of Dumouriez. But after the French had been driven back under the guns of Valenciennes, and were disorganized and unable to resist, why did the allies remain six months in front of a few towns and permit the Committee of Public Safety to organize new armies? When the deplorable condition of France and the destitution of the wreck of the army of Dampierre are considered, can the parades of the allies in front of the fortresses in Flanders be understood?

Invasions of a country whose strength lies mainly in the capital are particularly advantageous. Under the government of a powerful prince, and in ordinary wars, the most important point is the head-quarters of the army; but under a weak prince, in a republic, and still more in wars of opinion, the capital is generally the center of national power.[1] If this is ever doubtful, it was not so on this occasion. Paris was France, and this to such an extent that two-thirds of the nation had risen against the government which oppressed them. If, after having beaten the French army at Famars, the allies had left the Dutch and Hanoverians to observe what remained of it, while the English and the Austrians directed their operations upon the Meuse, the Sarre, and

---

[1]   The capture of Paris by the allies decided the fate of Napoleon; but he had no army, and was attacked by all Europe, and the French people had, in addition, separated their cause from his. If he had possessed fifty thousand more old soldiers, he would have shown that the capital was at his head-quarters.

the Moselle, in concert with the Prussians and a part of the useless army of the Upper Rhine, a force of one hundred and twenty thousand men, with its flanks protected by other troops, could have been pushed forward. It is even probable that, without changing the direction of the war or running great risks, the Dutch and Hanoverians could have performed the duty of observing Maubeuge and Valenciennes, while the bulk of the army pursued the remains of Dampierre's forces. After gaining several victories, however, two hundred thousand men were engaged in carrying on a few sieges and were not gaining a foot of ground. While they threatened France with invasion, they placed fifteen or sixteen bodies of troops, defensively, to cover their own frontier! When Valenciennes and Mayence capitulated, instead of falling with all their forces upon the camp at Cambray, they flew off, excentrically, to Dunkirk on one side and Landau on the other.

It is not less astonishing that, after making the greatest efforts in the beginning of the campaign upon the right of the general field, they should have shifted them afterward to the extreme left, so that while the allies were operating in Flanders they were in no manner seconded or aided by the imposing army upon the Rhine; and when, in its turn, this army took up the offensive, the allies remained inactive upon the Sambre. Do not these false combinations resemble those of Soubise and Broglie in 1761, and all the operations of the Seven Years' War?

In 1794 the phase of affairs is wholly changed. The French from a painful defensive pass to a brilliant offensive. The combinations of this campaign were doubtless well considered; but it is wrong to represent them as forming a new system of war. To be convinced of this, it is only necessary to observe that the respective positions of the armies in this campaign and in that of 1757 were almost identical, and the direction of the operations is quite the same. The French had four corps, which constituted two armies, as the King of Prussia had four divisions, which composed two armies.

These two large bodies took a concentric direction leading on Brussels, as Frederick and Schwerin had adopted in 1757 on Prague. The only difference between the two plans is that the Austrian troops in Flanders were not so much scattered as those of Brown in Bohemia; but this difference is certainly not favorable to the plan of 1794. The position of the North Sea was also unfavorable for the latter plan. To outflank the Austrian right, Pichegru was thrown between the sea and the mass of the enemy,—a direction as dangerous and faulty as could be given to great operations. This movement was the same as that of Benningsen on the Lower Vistula which almost lost the Russian army in 1807. The fate of the Prussian army, cut off from its communications and forced upon the Baltic, is another proof of this truth.

If the Prince of Coburg had acted with ability, he could easily have made Pichegru suffer for this audacious maneuver, which was performed a month before Jourdan was prepared to follow it up.

The center of the grand Austrian army intended to act upon the offensive was before Landrecies; the army was composed of one hundred and six battalions and one hundred and fifty squadrons; upon its right flank Flanders was covered by the corps d'armee of Clairfayt, and upon the left Charleroi was covered by that of the Prince de Kaunitz. The gain of a battle before Landrecies opened its gates; and upon General Chapuis was found a plan of the diversion in Flanders: only *twelve battalions* were sent to Clairfayt. A long time afterward, and after the French were known to have been successful, the corps of the Duke of York marched to Clairfayt's relief; but what was the use of the remainder of the army before Landrecies, after it was obliged by a loss of force to delay invasion? The Prince of Coburg threw away all the advantages of his central position, by allowing the French to concentrate in Belgium and to beat all his large detachments in detail.

Finally, the army moved, leaving a division at Cateau, and a part having been sent to the Prince de Kaunitz at Charleroi. If, instead of dividing this grand army, it had been directed upon Turcoing, there would have been concentrated there one hundred battalions and one hundred and forty squadrons; and what must then have been the result of this famous diversion of Pichegru, cut off from his own frontiers and shut up between the sea and two fortresses?

The plan of invasion adopted by the French had not only the radical error of exterior lines: it also failed in execution. The diversion on Courtray took place on April 26, and Jourdan did not arrive at Charleroi till the 3d of June,—more than a month afterward. Here was a splendid opportunity for the Austrians to profit by their central position. If the Prussian army had maneuvered by its right and the Austrian army by its left,—that is, both upon the Meuse,—the state of affairs would have been different. By establishing themselves in the center of a line of scattered forces they could have prevented the junction of the different fractions. It may be dangerous in a battle to attack the center of a close line of troops when it can be simultaneously sustained by the wings and the reserves; but it is quite different on a line of three hundred miles in extent.

In 1795 Prussia and Spain retired from the coalition, and the principal theater of war was shifted from the Rhine to Italy,—which opened a new field of glory for the French arms. Their lines of operations in this campaign were double; they desired to operate by Dusseldorf and Manheim. Clairfayt, wiser than his predecessors, concentrated his forces alternately upon these points, and gained victories at Manheim and in the lines of Mayence so decisive that they caused the army of the Sambre and Meuse to recross the Rhine to cover the Moselle, and brought Pichegru back to Landau.

In 1796 the lines of operations on the Rhine were copied from those of 1757 and those in Flanders in 1794, but with different results. The armies of the Rhine, and of the Sambre and Meuse, set out from the extremities of the base, on routes converging to the Danube. As in 1794, they were exterior lines. The Archduke Charles, more skillful than the Prince of Coburg, profited by his interior lines by concentrating his forces at a point nearer than that expected by the French. He then seized the instant when the Danube covered the corps of Latour, to steal several marches upon Moreau and attack and overwhelm Jourdan: the battle of Wurzburg decided the fate of Germany and compelled the army of Moreau to retreat.

Bonaparte now commences in Italy his extraordinary career. His plan is to separate the Piedmontese and Austrian armies. He succeeds by the battle of Millesimo in causing them to take two exterior strategic lines, and beats them successively at Mondovi and Lodi. A formidable army is collected in the Tyrol to raise the siege of Mantua: it commits the error of marching there in two bodies separated by a lake. The lightning is not quicker than Napoleon. He raises the siege, abandons every thing before Mantua, throws the greater part of his force upon the first column, which debouches by Brescia, beats it and forces it back upon the mountains: the second column arrives upon the same ground, and is there beaten in its turn, and compelled to retire into the Tyrol to keep up its communications with the right. Wurmser, upon whom these lessons are lost, desires to cover the two lines of Roveredo and Vicenza; Napoleon, after having overwhelmed and thrown the first back upon the Lavis, changes direction by the right, debouches by the gorges of the Brenta upon the left, and forces the remnant of this fine army to take refuge in Mantua, where it is finally compelled to surrender.

In 1799 hostilities recommence: the French, punished for having formed two exterior lines in 1796, nevertheless, have three upon the Rhine and the Danube. The army on the left observes the Lower Rhine, that of the center marches upon the Danube, Switzerland, flanking Italy and Swabia, being occupied by a third army as strong as both the others. *The three armies could be concentrated only in the valley of the Inn*, eighty leagues from their base of operations. The archduke has equal forces: he unites them against the center, which he defeats at Stockach, and the army of Switzerland is compelled to evacuate the Grisons and Eastern Switzerland. The allies in turn commit the same fault: instead of following up their success on this central line, which cost them so dearly afterward, they formed a double line in Switzerland and on the Lower Rhine. The army of Switzerland is beaten at Zurich, while the other trifles at Manheim.

In Italy the French undertake a double enterprise, which leaves thirty-two thousand men uselessly employed at Naples, while upon the Adige, where the vital blows were to be given or received, their force is too weak and meets with terrible reverses. When the army of Naples returns to the North, it commits the error of adopting a strategic direction opposed to Moreau's, and Suwaroff, by means of his central position, from which he derives full profit, marches against this army and beats it, while some leagues from the other.

In 1800, Napoleon has returned from Egypt, and every thing is again changed, and this campaign presents a new combination of lines of operations; one hundred and fifty thousand men march upon the two flanks of Switzerland,

and debouch, one upon the Danube and the other upon the Po. This insures the conquest of vast regions. Modern history affords no similar combination. The French armies are upon interior lines, affording reciprocal support, while the Austrians are compelled to adopt an exterior line, which renders it impossible for them to communicate. By a skillful arrangement of its progress, the army of the reserve cuts off the enemy from his line of operations, at the same time preserving its own relations with its base and with the army of the Rhine, which forms its secondary line.

Fig. 3 demonstrates this truth, and shows the respective situations of the two parties. A and A A indicate the front of operations of the armies of the Rhine and of the reserve; B and B B, that of Kray and Melas; C C C C, the passes of the Saint-Bernard, of the Simplon, of the Saint-Gothard, and of the Splugen; D indicates the two lines of operations of the army of the reserve; E, the two lines of retreat of Melas; H J K, the French divisions preserving their line of retreat. It may thus be seen that Melas is cut off from his base, and that, on the contrary, the French general runs no risk, since he preserves all his communications with the frontiers and with his secondary lines.

The analysis of the memorable events just sketched shows clearly the importance of a proper selection of lines of maneuver in military operations. Indeed, discretion on this point may repair the disasters of defeat, destroy the advantages of an adversary's victory, render his invasion futile, or assure the conquest of a province.

By a comparison of the combinations and results of the most noted campaigns, it will be seen that the lines of operations which have led to success have been established in conformity to the fundamental principle already alluded to,— viz.: that *simple and interior lines enable a general to bring into action, by strategic movements, upon the important point, a stronger force than the enemy.* The student may also satisfy himself that those which have failed contained faults opposed to this principle. An undue number of lines divides the forces, and permits fractions to be overwhelmed by the enemy.

Figure 3. The Stratigic Field of 1806
To illustrtate Maxim 3 on the Direction of Lines of Operations

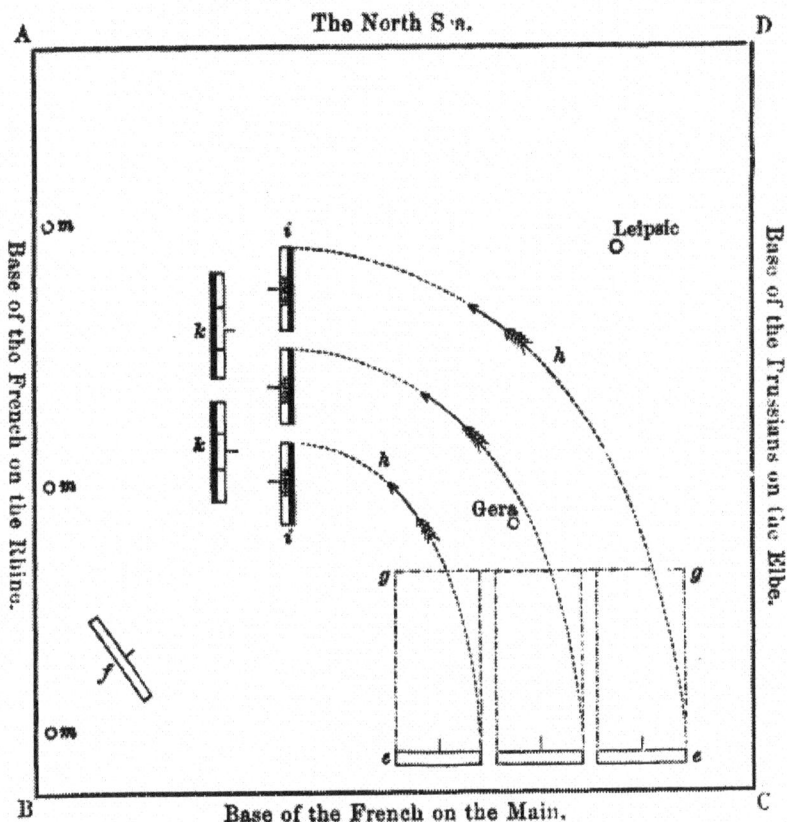

The North Sa.

A   D

Base of the French on the Rhine.

Base of the Prussians on the Elbe.

Leipsic

Gera

Base of the French on the Main.

B   C

MAXIMS ON LINES OF OPERATIONS.

From the analysis of all the events herein referred to, as well as from that of many others, the following maxims result:–

1. If the art of war consists in bringing into action upon the decisive point of the theater of operations the greatest possible force, the choice of the line of operations, being the primary means of attaining this end, may be regarded as the fundamental idea in a good plan of a campaign. Napoleon proved this by the direction he gave his armies in 1805 on Donauwerth and in 1806 on Gera,—maneuvers that cannot be too much studied by military men.

Of course, it is impossible to sketch in advance the whole campaign. The objective point will be determined upon in advance, the general plan to be followed to attain it, and the first enterprise to be undertaken for this end: what is to follow will depend upon the result of this first operation and the new phases it may develop.

2. The direction to be given to this line depends upon the geographical situation of the theater of operations, but still more upon the position of the hostile masses upon this strategic field. *In every case, however, it must be directed upon the center or upon one of the extremities. Only when the assailing forces are vastly preponderating would it be otherwise than a fatal error to act upon the center and the two extremities at the same time.*[1]

It may be laid down as a general principle, that, if the enemy divide his forces on an extended front, the best direction of the maneuver-line will be upon his center, but in every other case, when it is possible, the best direction will be upon one of the flanks, and then upon the rear of his line of defense or front of operations.

The advantage of this maneuver arises more from the opportunity it affords of taking the line of defense in reverse than from the fact that by using it the assailant has to contend with but a part of the enemy's force. Thus, the army of the Rhine in 1800, gaining the extreme left of the line of defense of the Black Forest, caused it to yield almost without an effort. This army fought two battles on the right bank of the Danube, which, although not decisive, yet, from the judicious direction of the line of operations, brought about the invasion of Swabia and Bavaria. The results of the march of the army of the reserve by the Saint-Bernard and Milan upon the extreme right of Melas were still more brilliant.

3. Even when the extremity of the enemy's front of operations is gained, it is not always safe to act upon his rear, since by so doing the assailant in many cases will lose his own communications. To avoid this danger, the line of operations should have a geographic and strategic direction, such that the army will always find either to its rear or to the right or left a safe line of retreat. In this case, to take advantage of either of these flank lines of retreat would require a change of direction of the line of operations, (Maxim 12.)

The ability to decide upon such a direction is among the most important qualities of a general. The importance of a direction is illustrated by these examples.

If Napoleon in 1800, after passing the Saint-Bernard, had marched upon Asti or Alessandria, and had fought at Marengo without having previously protected himself on the side of Lombardy and of the left bank of the Po, he would have been more thoroughly cut off from his line of retreat than Melas from his; but, having in his possession the secondary points of Casale and Pavia on the side of the Saint-Bernard, and Savona and Tenda toward the Apennines, in case of reverse he had every means of regaining the Var or the Valais.

---

[1]   The inferiority of an army does not depend exclusively upon the number of soldiers: their military qualities, their *morale*, and the ability of their commander are also very important elements.

In 1806, if he had marched from Gera directly upon Leipsic, and had there awaited the Prussian army returning from Weimar, he would have been cut off from the Rhine as much as the Duke of Brunswick from the Elbe, while by falling back to the west in the direction of Weimar he placed his front before the three roads of Saalfeld, Schleiz, and Hof, which thus became well-covered lines of communication. If the Prussians had endeavored to cut him off from these lines by moving between Gera and Baireuth, they would have opened to him his most natural line,—the excellent road from Leipsic to Frankfort,—as well as the two roads which lead from Saxony by Cassel to Coblentz, Cologne, and even Wesel.

4. Two independent armies should not be formed upon the same frontier: such an arrangement could be proper only in the case of large coalitions, or where the forces at disposal are too numerous to act upon the same zone of operations; and even in this case it would be better to have all the forces under the same commander, who accompanies the principal army.

5. As a consequence of the last-mentioned principle, with equal forces on the same frontier, a single line of operations will be more advantageous than a double one.

6. It may happen, however, that a double line will be necessary, either from the topography of the seat of war, or because a double line has been adopted by the enemy, and it will be necessary to oppose a part of the army to each of his masses.

7. In this case, interior or central lines will be preferable to exterior lines, since in the former case the fractions of the army can be concentrated before those of the enemy, and may thus decide the fate of the campaign.[1] Such an army may, by a well-combined strategic plan, unite upon and overwhelm successively the fractions of the adversary's forces. To be assured of success in these maneuvers, a body of observation is left in front of the army to be held in check, with instructions to avoid a serious engagement, but to delay the enemy as much as possible by taking advantage of the ground, continually falling back upon the principal army.

8. A double line is applicable in the case of a decided superiority of force, when each army will be a match for any force the enemy can bring against it. In this case this course will be advantageous,—since a single line would crowd the forces so much as to prevent them all from acting to advantage. However, it will always be prudent to support well the army which, by reason of the nature of its theater and the respective positions of the parties, has the most important duty to perform.

9 The principal events of modern wars demonstrate the truth of two other maxims. The first is, that two armies operating on interior lines and sustaining each other reciprocally, and opposing two armies superior in numbers, should

---

1   When the fractions of an army are separated from the main body by only a few marches, and particularly when they are not intended to act separately throughout the campaign, these are central strategic positions, and not lines of operations.

not allow themselves to be crowded into a too contracted space, where the whole might be overwhelmed at once. This happened to Napoleon at Leipsic.[1] The second is, that interior lines should not be abused by extending them too far, thus giving the enemy the opportunity of overcoming the corps of observation. This risk, however, may be incurred if the end pursued by the main forces is so decisive as to conclude the war,—when the fate of these secondary bodies would be viewed with comparative indifference.

10. For the same reason, two converging lines are more advantageous than two divergent. The first conform better to the principles of strategy, and possess the advantage of covering the lines of communication and supply; but to be free from danger they should be so arranged that the armies which pass over them shall not be separately exposed to the combined masses of the enemy, before being able to effect their junction.

11. Divergent lines, however, may be advantageous when the center of the enemy has been broken and his forces separated either by a battle or by a strategic movement,—in which case divergent operations would add to the dispersion of the enemy. Such divergent lines would be interior, since the pursuers could concentrate with more facility than the pursued.

12. It sometimes happens that an army is obliged to change its line of operations in the middle of a campaign. This is a very delicate and important step, which may lead to great successes, or to equally great disasters if not applied with sagacity, and is used only to extricate an army from an embarrassing position. Napoleon projected several of these changes; for in his bold invasions he was provided with new plans to meet unforeseen events.

At the battle of Austerlitz, if defeated, he had resolved to adopt a line of operations through Bohemia on Passau or Ratisbon, which would have opened a new and rich country to him, instead of returning by Vienna, which route lay through an exhausted country and from which the Archduke Charles was endeavoring to cut him off. Frederick executed one of these changes of the line of operations after the raising of the siege of Olmutz.

In 1814 Napoleon commenced the execution of a bolder maneuver, but one which was favored by the localities. It was to base himself upon the fortresses of Alsace and Lorraine, leaving the route to Paris open to the allies. If Mortier and Marmont could have joined him, and had he possessed fifty thousand more men, this plan would have produced the most decisive results and have put the seal on his military career.

13. As before stated, the outline of the frontiers, and the geographical character of the theater of operations, exercise a great influence on the direction to be given to these lines, as well as upon the advantages to be obtained. Central positions, salient toward the enemy, like Bohemia and Switzerland, are the most advantageous, because they naturally lead to the adoption of interior

---

[1] In the movements immediately preceding the battle of Leipsic, Napoleon, strictly speaking, had but a single line of operations, and his armies were simply in central strategic positions; but the principle is the same, and hence the example is illustrative of lines of operations.

lines and facilitate the project of taking the enemy in reverse. The sides of this salient angle become so important that every means should be taken to render them impregnable. In default of such central positions, their advantages may be gained by the relative directions of maneuver-lines, as the following figure will explain. C D maneuvering upon the right of the front of the army A B, and H I upon the left flank of G F, will form two interior lines I K and C K upon an extremity of the exterior lines A B, F G, which they may overwhelm separately by combining upon them. Such was the result of the operations of 1796, 1800, and 1809.

Figure 4

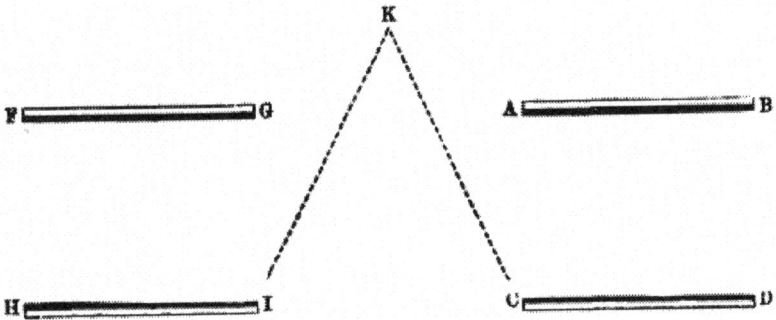

14. The general configuration of the bases ought also to influence the direction to be given to the lines of operations, these latter being naturally dependent upon the former. It has already been shown that the greatest advantage that can result from a choice of bases is when the frontiers allow it to be assumed parallel to the line of operations of the enemy, thus affording the opportunity of seizing this line and cutting him from his base.

But if, instead of directing the operations upon the decisive point, the line of operations be badly chosen, all the advantages of the perpendicular base may be lost, as will be seen by referring to the figure. The army E, having the double base A C and C D, if it marched toward F, instead of to the right toward G H, would lose all the strategic advantages of its base C D.

The great art, then, of properly directing lines of operations, is so to establish them in reference to the bases and to the marches of the army as to seize the communications of the enemy without imperiling one's own, and is the most important and most difficult problem in strategy.

15. There is another point which exercises a manifest influence over the direction to be given to the line of operations; it is when the principal enterprise of the campaign is to cross a large river in the presence of a numerous and well-appointed enemy. In this case, the choice of this line depends neither upon the will of the general nor the advantages to be gained by an attack on one or

another point; for the first consideration will be to ascertain where the passage can be most certainly effected, and where are to be found the means for this purpose. The passage of the Rhine in 1795, by Jourdan, was near Dusseldorf, for the same reason that the Vistula in 1831 was crossed by Marshal Paskevitch near Ossiek,—viz., that in neither case was there the bridge-train necessary for the purpose, and both were obliged to procure and take up the rivers large boats, bought by the French in Holland, and by the Russians at Thorn and Dantzic. The neutrality of Prussia permitted the ascent of the river in both cases, and the enemy was not able to prevent it. This apparently incalculable advantage led the French into the double invasions of 1795 and 1796, which failed because the double line of operations caused the defeat of the armies separately. Paskevitch was wiser, and passed the Upper Vistula with only a small detachment and after the principal army had already arrived at Lowicz.

When an army is sufficiently provided with bridge-trains, the chances of failure are much lessened; but then, as always, it is necessary to select the point which may, either on account of its topography or the position of the enemy, be most advantageous. The discussion between Napoleon and Moreau on the passage of the Rhine in 1800 is one of the most curious examples of the different combinations presented by this question, which is both strategic and tactical.

Since it is necessary to protect the bridges, at least until a victory is gained, the point of passage will exercise an influence upon the directions of a few marches immediately subsequent to the passage. The point selected in every case for the principal passage will be upon the center or one of the flanks of the enemy.

A united army which has forced a passage upon the center of an extended line might afterward adopt two divergent lines to complete the dispersion of the enemy, who, being unable to concentrate, would not think of disturbing the bridges.

If the line of the river is so short that the hostile army is more concentrated, and the general has the means of taking up after the passage a front perpendicular to the river, it would be better to pass it upon one of the extremities, in order to throw off the enemy from the bridges. This will be referred to in the article upon the passage of rivers.

16. There is yet another combination of lines of operations to be noticed. It is the marked difference of advantage between a line at home and one in a hostile country. The nature of the enemy's country will also influence these chances. Let us suppose an army crosses the Alps or the Rhine to carry on war in Italy or Germany. It encounters states of the second rank; and, even if they are in alliance, there are always rivalries or collisions of interest which will deprive them of that unity and strength possessed by a single powerful state. On the other hand, a German army invading France would operate upon a line much more dangerous than that of the French in Italy, because upon the first could be thrown the consolidated strength of Franco, united in feeling

and interest. An army on the defensive, with its line of operations on its own soil, has resources everywhere and in every thing: the inhabitants, authorities, productions, towns, public depots and arsenals, and even private stores, are all in its favor. It is not ordinarily so abroad.

Lines of operations in rich, fertile, manufacturing regions offer to the assailants much greater advantages than when in barren or desert regions, particularly when the people are not united against the invader. In provinces like those first named the army would find a thousand necessary supplies, while in the other huts and straw are about the only resources. Horses probably may obtain pasturage; but every thing else must be carried by the army,—thus infinitely increasing the embarrassments and rendering bold operations much more rare and dangerous. The French armies, so long accustomed to the comforts of Swabia and Lombardy, almost perished in 1806 in the bogs of Pultusk, and actually did perish in 1812 in the marshy forests of Lithuania.

17. There is another point in reference to these lines which is much insisted upon by some, but which is more specious than important. It is that on each side of the line of operations the country should be cleared of all enemies for a distance equal to the depth of this line: otherwise the enemy might threaten the line of retreat. This rule is everywhere belied by the events of war. The nature of the country, the rivers and mountains, the morale of the armies, the spirit of the people, the ability and energy of the commanders, cannot be estimated by diagrams on paper. It is true that no considerable bodies of the enemy could be permitted on the flanks of the line of retreat; but a compliance with this demand would deprive an army of every means of taking a step in a hostile country; and there is not a campaign in recent wars, or in those of Marlborough and Eugene, which does not contradict this assertion. Was not General Moreau at the gates of Vienna when Fussen, Scharnitz, and all the Tyrol were in possession of the Austrians? Was not Napoleon at Piacenza when Turin, Genoa, and the Col-di-Tenda were occupied by the army of Melas? Did not Eugene march by way of Stradella and Asti to the aid of Turin, leaving the French upon the Mincio but a few leagues from his base?

OBSERVATIONS UPON INTERIOR LINES—WHAT HAS BEEN SAID AGAINST THEM

Some of my critics have disputed as to the meaning of words and upon definitions; others have censured where they but imperfectly understood; and others have, by the light of certain important events, taken it upon themselves to deny my fundamental principles, without inquiring whether the conditions of the case which might modify the application of these principles were such as were supposed, or without reflecting that, even admitting what they claimed to be true, a single exception cannot disprove a rule based upon the experience of ages and upon natural principles.

In opposition to my maxims upon interior lines, some have quoted the famous and successful march of the allies upon Leipsic. This remarkable event, at first glance, seems to stagger the faith of those who believe in principles. At best, however, it is but one of those exceptional cases from which nothing can be inferred in the face of thousands of opposed instances. Moreover, it is easy to show that, far from overthrowing the maxims it has been brought to oppose, it will go to establish their soundness. Indeed, the critics had forgotten that in case of a considerable numerical superiority I recommended double lines of operations as most advantageous, particularly when concentric and arranged to combine an effort against the enemy at the decisive moment. Now, in the allied armies of Schwarzenberg, Bluecher, Bernadotte, and Benningsen, this case of decided superiority is found. The inferior army, to conform to the principles of this chapter, should have directed its efforts against one of the extremities of his adversary, and not upon the center as it did: so that the events quoted against me are doubly in my favor.

Moreover, if the central position of Napoleon between Dresden and the Oder was disastrous, it must be attributed to the misfortunes of Culm, Katzbach, and Dennewitz,—in a word, to faults of execution, entirely foreign to the principles in question.

What I propose is, to act offensively upon the most important point with the greater part of the forces, but upon the secondary points to remain on the defensive, in strong positions or behind a river, until the decisive blow is struck, and the operation ended by the total defeat of an essential part of the army. Then the combined efforts of the whole army may be directed upon other points. Whenever the secondary armies are exposed to a decisive shock during the absence of the mass of the army, the system is not understood; and this was what happened in 1813.

If Napoleon, after his victory at Dresden, had vigorously pursued the allies into Bohemia, he would have escaped the disaster at Culm, have threatened Prague, and perhaps have dissolved the Coalition. To this error may be added a fault quite as great,—that of fighting decisive battles when he was not present with the mass of his forces. At Katzbach his instructions were not obeyed. He ordered Macdonald to wait for Bluecher, and to fall upon him when he should expose himself by bold movements. Macdonald, on the contrary, crossed his detachments over torrents which were hourly becoming more swollen, and advanced to meet Bluecher. If he had fulfilled his instructions and Napoleon had followed up his victory, there is no doubt that his plan of operations, based upon interior strategic lines and positions and upon a concentric line of operations, would have met with the most brilliant success. The study of his campaigns in Italy in 1796 and in France in 1814 shows that he knew how to apply this system.

There is another circumstance, of equal importance, which shows the injustice of judging central lines by the fate of Napoleon in Saxony,—viz.: *that his front of operations was outflanked on the right, and even taken in reverse, by the geographical position of the frontiers of Bohemia.* Such a case is of rare occurrence. A central position with such faults is not to be compared to one without them. When Napoleon made the application of these principles in Italy, Poland, Prussia, and France, he was not exposed to the attack of a hostile enemy on his flanks and rear. Austria could have threatened him in 1807; but she was then at peace with him and unarmed. To judge of a system of operations, it must be supposed that accidents and chances are to be as much in favor of as against it,—which was by no means the case in 1813, either in the geographic positions or in the state of the respective forces. Independently of this, it is absurd to quote the reverses at Katzbach and Dennewitz, suffered by his lieutenants, as proof capable of destroying a principle the simplest application of which required these officers not to allow themselves to be drawn into a serious engagement. Instead of avoiding they sought collisions. Indeed, what advantage can be expected from the system of central lines, if the parts of the army which have been weakened in order to strike decisive blows elsewhere, shall themselves seek a disastrous contest, instead of being contented with being bodies of observation?[1] In this case it is the enemy who applies the principle, and not he who has the interior lines. Moreover, in the succeeding campaign, the defense of Napoleon in Champagne, from the battle of Brienne to that of Paris, demonstrates fully the truth of these maxims.

The analysis of these two celebrated campaigns raises a strategic question which it would be difficult to answer by simple assertions founded upon theories. It is, whether the system of central lines loses its advantages when the masses are very large. Agreeing with Montesquieu, that the greatest enterprises fail from the magnitude of the arrangements necessary to consummate them, I am disposed to answer in the affirmative. It is very clear to me that an army of one hundred thousand men, occupying a central zone against three isolated armies of thirty or thirty-five thousand men, would be more sure of defeating them successively than if the central mass were four hundred thousand strong against three armies of one hundred and thirty-five thousand each; and for several good reasons:–

1. Considering the difficulty of finding ground and time necessary to bring a very large force into action on the day of battle, an army of one hundred and thirty or one hundred and forty thousand men may easily resist a much larger force.

2. If driven from the field, there will be at least one hundred thousand men to protect and insure an orderly retreat and effect a junction with one of the other armies.

3. The central army of four hundred thousand men requires such a quantity of provisions, munitions, horses, and *materiel* of every kind, that it will possess less mobility and facility in shifting its efforts from one part of

---

[1]   I am well aware that it is not always possible to avoid a combat without running greater risks than would result from a check; but Macdonald might have fought Bluecher to advantage if he had better understood Napoleon's instructions.

the zone to another; to say nothing of the impossibility of obtaining provisions from a region too restricted to support such numbers.

4. The bodies of observation detached from the central mass to hold in check two armies of one hundred and thirty-five thousand each must be very strong, (from eighty to ninety thousand each;) and, being of such magnitude, if they are drawn into a serious engagement they will probably suffer reverses, the effects of which might outweigh the advantages gained by the principal army.

I have never advocated exclusively either a concentric or eccentric system. All my works go to show the eternal influence of principles, and to demonstrate that operations to be successful must be applications of principles.

Divergent or convergent operations may be either very good or very bad: all depends on the situation of the respective forces. The eccentric lines, for instance, are good when applied to a mass starting from a given point, and acting in divergent directions to divide and separately destroy two hostile forces acting upon exterior lines. Such was the maneuver of Frederick which brought about, at the end of the campaign of 1767, the fine battles of Rossbach and Leuthen. Such were nearly all the operations of Napoleon, whose favorite maneuver was to unite, by closely-calculated marches, imposing masses on the center, and, having pierced the enemy's center or turned his front, to give them eccentric directions to disperse the defeated army.[1]

On the other hand, concentric operations are good in two cases: 1. When they tend to concentrate a scattered army upon a point where it will be sure to arrive before the enemy; 2. When they direct to the same end the efforts of two armies which are in no danger of being beaten separately by a stronger enemy.

Concentric operations, which just now seem to be so advantageous, may be most pernicious,—which should teach us the necessity of detecting the principles upon which systems are based, and not to confound principles and systems; as, for instance, if two armies set out from a distant base to march convergently upon an enemy whose forces are on interior lines and more concentrated, it follows that the latter could effect a union before the former, and would inevitably defeat them; as was the case with Moreau and Jourdan in 1796, opposed to the Archduke Charles.

In starting from the same points, or from two points much less separated than Dusseldorf and Strasbourg, an army may be exposed to this danger. What was the fate of the concentric columns of Wurmser and Quasdanovitch, wishing to reach the Mincio by the two banks of Lake Garda? Can the result of the march of Napoleon and Grouchy on Brussels be forgotten? Leaving Sombref, they were to march concentrically on this city,—one by Quatre-Bras,

---

[1]   It will not be thought strange that I sometimes approve of concentric, and at other times divergent, maneuvers, when we reflect that among the finest operations of Napoleon there are some in which he employed these two systems alternately within twenty-four hours; for example, in the movements about Ratisbon in 1809.

the other by Wavre. Bluecher and Wellington, taking an interior strategic line, effected a junction before them, and the terrible disaster of Waterloo proved to the world that the immutable principles of war cannot be violated with impunity.

Such events prove better than any arguments that a system which is not in accordance with the principles of war cannot be good. I lay no claim to the creation of these principles, for they have always existed, and were applied by Caesar, Scipio, and the Consul Nero, as well as by Marlborough and Eugene; but I claim to have been the first to point them out, and to lay down the principal chances in their various applications.

ARTICLE XXII

STRATEGIC LINES

Mention has already been made of strategic lines of maneuvers, which differ essentially from lines of operations; and it will be well to define them, for many confound them. We will not consider those strategic lines which have a great and permanent importance by reason of their position and their relation to the features of the country, like the lines of the Danube and the Meuse, the chains of the Alps and the Balkan. Such lines can best be studied by a detailed and minute examination of the topography of Europe; and an excellent model for this kind of study is found in the Archduke Charles's description of Southern Germany.

The term *strategic* is also applied to all communications which lead by the most direct or advantageous route from one important point to another, as well as from the strategic front of the army to all of its objective points. It will be seen, then, that a theater of war is crossed by a multitude of such lines, but that at any given time those only which are concerned in the projected enterprise have any real importance. This renders plain the distinction between the general line of operations of a whole campaign, and these *strategic* lines, which are temporary and change with the operations of the army.

Besides territorial strategic lines, there are *strategic lines of maneuvers*.

An army having Germany as its general field might adopt as its zone of operations the space between the Alps and the Danube, or that between the Danube and the Main, or that between the mountains of Franconia and the sea. It would have upon its zone a single line of operations, or, at most, a double concentric line, upon interior, or perhaps exterior, directions,—while it would have successively perhaps twenty strategic lines as its enterprises were developed: it would have at first one for each wing which would join the general line of operations. If it operated in the zone between the Danube and the Alps, it might adopt, according to events, the strategic line leading from

Ulm on Donauwerth and Ratisbon, or that from Ulm to the Tyrol, or that which connects Ulm with Nuremberg or Mayence.

It may, then, be assumed that the definitions applied to lines of operations, as well as the maxims referring to them, are necessarily applicable to strategic lines. These may be *concentric*, to inflict a decisive blow, or *eccentric*, after victory. They are rarely *simple*, since an army does not confine its march to a single road; but when they are double or triple, or even quadruple, they should be *interior* if the forces be equal, or *exterior* in the case of great numerical superiority. The rigorous application of this rule may perhaps sometimes be remitted in detaching a body on an exterior line, even when the forces are equal, to attain an important result without running much risk; but this is an affair of detachments, and does not refer to the important masses.

Strategic lines cannot be interior when our efforts are directed against one of the extremities of the enemy's front of operations.

The maxims above given in reference to lines of operations holding good for strategic lines, it is not necessary to repeat them, or to apply them to particular examples; but there is one, however, which deserves mention,—viz.: that it is important generally, in the selection of these temporary strategic lines, not to leave the line of operations exposed to the assaults of the enemy. Even this may, however, be done, to extricate the army from great danger, or to attain a great success; but the operation must be of short duration, and care must have been taken to prepare a plan of safe retreat, by a sudden change of the line of operations, if necessary, as has already been referred to.

We will illustrate this by the campaign of Waterloo. The Prussian army was based upon the Rhine, its line of operations extended from Cologne and Coblentz on Luxembourg and Namur; Wellington's base was Antwerp, and his line of operations the short road to Brussels. The sudden attack by Napoleon on Flanders decided Bluecher to receive battle parallel to the English base, and not to his own, about which he seemed to have no uneasiness. This was pardonable, because he could always have a good chance of regaining Wesel or Nimeguen, and even might seek a refuge in Antwerp in the last extremity; but if the army had not had its powerful maritime allies it would have been destroyed. Beaten at Ligny, and seeking refuge at Gembloux and then at Wavre, Bluecher had but three strategic lines to choose from: that which led directly to Maestricht, that farther north on Venloo, or the one leading to the English army near Mont St. Jean. He audaciously took the last, and triumphed by the application of interior strategic lines,—which Napoleon here, perhaps for the first time in his life, neglected. It will readily be seen that the line followed from Gembloux by Wavre to Mont St. Jean was neither a line of operations of the Prussian army nor a line of battle, but a *strategic line of maneuver*, and was interior. It was bold, because he exposed fully his own natural line of operations. The fact that he sought a junction with the English made his movement accord with the principles of war.

A less successful example was that of Ney at Dennewitz. Leaving Wittenberg, and going in the direction of Berlin, he moved to the right to gain the extreme left of the allies, but in so doing he left his primitive line of retreat exposed to the attacks of an enemy superior in force. His object was to gain communication with Napoleon, whose intention was to join him by Herzberg or Luckau; but Ney should from the beginning have taken all logistic and tactical means of accomplishing this change of strategic line and of informing his army of it. He did nothing of this kind,—either from forgetfulness, or on account of the feeling of aversion he had to any thing like a retreat,—and the severe losses at Dennewitz were the result.

Napoleon in 1796 gave one of the best illustrations of these different combinations of strategic lines. His general line of operations extended from the Apennines to Verona. When he had driven Wurmser upon Roveredo and determined to pursue him into the Tyrol, he pushed on in the valley of the Adige to Trent and the Lavis, where he learned that Wurmser had moved by the Brenta on the Frioul, doubtless to take him in reverse. There were but three courses open to him,—to remain in the narrow valley of the Adige at great risk, to retreat by Verona to meet Wurmser, or the last,—which was sublime, but rash,—to follow him into the valley of the Brenta, which was encircled by rugged mountains whose two passages might be held by the Austrians. Napoleon was not the man to hesitate between three such alternatives. He left Vaubois on the Lavis to cover Trent, and marched with the remainder of his forces on Bassano. The brilliant results of this bold step are well known. The route from Trent to Bassano was not the line of operations of the army, but a *strategic line of maneuver* still bolder than that of Bluecher on Wavre. However, it was an operation of only three or four days' duration, at the end of which time Napoleon would either beat or be beaten at Bassano: in the first case, he would open direct communication with Verona and his line of operations; in the second, he could regain in great haste Trent, where, reinforced by Vaubois, he could fall back either upon Verona or Peschiera. The difficulties of the country, which made this march audacious in one respect, were favorable in another; for even if Wurmser had been victorious at Bassano he could not have interfered with the return to Trent, as there was no road to enable him to anticipate Napoleon. If Davidovitch on the Lavis had driven Vaubois from Trent, he might have embarrassed Napoleon; but this Austrian general, previously beaten at Roveredo, and ignorant of what the French army was doing for several days, and thinking it was all upon him, would scarcely have thought of resuming the offensive before Napoleon beaten at Bassano would have been on his retreat. Indeed, if Davidovitch had advanced as far as Roveredo, driving Vaubois before him, he would there have been surrounded by two French armies, who would have inflicted upon him the fate of Vandamme at Culm.

I have dwelt on this event to show that a proper calculation of time and distances, joined to great activity, may lead to the success of many adventures which may seem very imprudent. I conclude from this that it may be well sometimes to direct an army upon a route which exposes its line of operations, but that every measure must be taken to prevent the enemy from profiting by it, both by great rapidity of execution and by demonstrations which will deceive him and leave him in ignorance of what is taking place. Still, it is a very hazardous maneuver, and only to be adopted under an urgent necessity.

ARTICLE XXIII

MEANS OF PROTECTING A LINE OF OPERATIONS BY TEMPORARY BASES OR STRATEGIC RESERVES

When a general enters a country offensively, he should form eventual or temporary bases,—which, of course, are neither so safe nor so strong as his own frontiers. A river with *tetes de ponts*, and one or two large towns secure from a *coup de main* to cover the depots of the army and to serve as points of assembling for the reserve troops, would be an excellent base of this kind. Of course, such a line could not be a temporary base if a hostile force were near the line of operations leading to the real base on the frontiers. Napoleon would have had a good real base on the Elbe in 1813 if Austria had remained neutral; but, she having joined his enemies, this line was taken in reverse, and became but a pivot of operations, favorable indeed for the execution of a single enterprise, but dangerous for a prolonged occupation, particularly in case of a serious reverse. As every army which is beaten in an enemy's country is exposed to the danger of being cut off from its own frontiers if it continues to occupy the country, these distant temporary bases are rather temporary points of support than real bases, and are in a measure eventual lines of defense. In general, we cannot expect to find in an enemy's country safe positions suitable even for a temporary base; and the deficiency must be supplied by a strategic reserve,—which is purely a modern invention. Its merits and demerits deserve notice.

STRATEGIC RESERVES

Reserves play an important part in modern warfare. From the executive, who prepares national reserves, down to the chief of a platoon of skirmishers, every commander now desires a reserve. A wise government always provides good reserves for its armies, and the general uses them when they come under his command. The state has its reserves, the army has its own, and every corps d'armee or division should not fail to provide one.

The reserves of an army are of two kinds,—those on the battle-field, and those which are intended to recruit and support the army: the latter, while organizing, may occupy important points of the theater of war, and serve even as strategic reserves; their positions will depend not only on their magnitude, but also on the nature of the frontiers and the distance from the base to the front of operations. Whenever an army takes the offensive, it should always contemplate the possibility of being compelled to act on the defensive, and by the posting of a reserve between the base and front of operations the advantage of an active reserve on the field of battle is gained: it can fly to the support of menaced points without weakening the active army. It is true that to form a reserve a number of regiments must be withdrawn from active service; but there are always reinforcements to arrive, recruits to be instructed, and convalescents to be used; and by organizing central depots for preparation of munitions and equipments, and by making them the rendezvous of all detachments going to and coming from the army, and adding to them a few good regiments to give tone, a reserve may be formed capable of important service.

Napoleon never failed to organize these reserves in his campaigns. Even in 1797, in his bold march on the Noric Alps, he had first Joubert on the Adige, afterward Victor (returning from the Roman States) in the neighborhood of Verona. In 1805 Ney and Augereau played the part alternately in the Tyrol and Bavaria, and Mortier and Marmont near Vienna.

In 1806 Napoleon formed like reserves on the Rhine, and Mortier used them to reduce Hesse. At the same time, other reserves were forming at Mayence under Kellermann, which took post, as fast as organized, between the Rhine and Elbe, while Mortier was sent into Pomerania. When Napoleon decided to push on to the Vistula in the same year, he directed, with much ostentation, the concentration of an army on the Elbe sixty thousand strong, its object being to protect Hamburg against the English and to influence Austria, whose disposition was as manifest as her interests.

The Prussians established a similar reserve in 1806 at Halle, but it was badly posted: if it had been established upon the Elbe at Wittenberg or Dessau, and had done its duty, it might have saved the army by giving Prince Hohenlohe and Bluecher time to reach Berlin, or at least Stettin.

These reserves are particularly useful when the configuration of the country leads to double fronts of operations: they then fulfill the double object of observing the second front, and, in case of necessity, of aiding the operations of the main army when the enemy threatens its flanks or a reverse compels it to fall back toward this reserve.

Of course, care must be taken not to create dangerous detachments, and whenever these reserves can be dispensed with, it should be done, or the troops in the depots only be employed as reserves. It is only in distant invasions and sometimes on our own soil that they are useful: if the scene of hostilities be but five or six marches distant from the frontier, they are quite superfluous. At home they may generally be dispensed with: it is only in the case of a serious invasion, when new levies are organizing, that such a reserve, in an intrenched camp, under the protection of a fortress which serves as a great depot, will be indispensable.

The general's talents will be exercised in judging of the use of these reserves according to the state of the country, the length of the line of operations, the nature of the fortified points, and the proximity of a hostile state. He also decides upon their position, and endeavors to use for this purpose troops which will not weaken his main army so much as the withdrawal of his good troops.

These reserves ought to hold the most important points between the base and front of operations, occupy the fortified places if any have been reduced, observe or invest those which are held by the enemy; and if there be no fortress as a point of support, they should throw up intrenched camps or *tetes de ponts* to protect the depots and to increase the strength of their positions.

All that has been said upon pivots of operations is applicable to temporary bases and to strategic reserves, which will be doubly valuable if they possess such well-located pivots.

ARTICLE XXIV

THE OLD SYSTEM OF WARS OF POSITION AND THE MODERN SYSTEM OF MARCHES

By the system of positions is understood the old manner of conducting a methodical war, with armies in tents, with their supplies at hand, engaged in watching each other; one besieging a city, the other covering it; one, perhaps, endeavoring to acquire a small province, the other counteracting its efforts by occupying strong points. Such was war from the Middle Ages to the era of the French Revolution. During this revolution great changes transpired, and many systems of more or less value sprang up. War was commenced in 1792 as it had been in 1762: the French encamped near their strong places, and the allies besieged them. It was not till 1793, when assailed from without and within, that this system was changed. Thoroughly aroused, France threw one million men in fourteen armies upon her enemies. These armies had neither tents, provisions, nor money. On their marches they bivouacked or were quartered in towns; their mobility was increased and became a means of success. Their tactics changed also: the troops were put in columns, which were more easily handled than deployed lines, and, on account of the broken character of the country of Flanders and the Vosges, they threw out a part of their force as skirmishers to protect and cover the columns. This system, which was thus the result of circumstances, at first met with a success beyond all expectation: it disconcerted the methodical Austrian and Prussian troops as well as their generals. Mack, to whom was attributed the success of the Prince of Coburg, increased his reputation by directing the troops to extend their lines to oppose an open order to the fire of skirmishers. It had never occurred to the poor man that while the skirmishers made the noise the columns carried the positions.

98

The first generals of the Republic were fighting-men, and nothing more. The principal direction of affairs was in the hands of Carnot and of the Committee of Public Safety: it was sometimes judicious, but often bad. Carnot was the author of one of the finest strategic movements of the war. In 1793 he sent a reserve of fine troops successively to the aid of Dunkirk, Maubeuge, and Landau, so that this small force, moving rapidly from point to point, and aided by the troops already collected at these different points, compelled the enemy to evacuate France.

The campaign of 1794 opened badly. It was the force of circumstances, and not a premeditated plan, which brought about the strategic movement of the army of the Moselle on the Sambre; and it was this which led to the success of Fleurus and the conquest of Belgium.

In 1795 the mistakes of the French were so great that they were imputed to treachery. The Austrians, on the contrary, were better commanded by Clairfayt, Chateler, and Schmidt than they had been by Mack and the Prince of Coburg. The Archduke Charles, applying the principle of interior lines, triumphed over Moreau and Jourdan in 1796 by a single march.

Up to this time the fronts of the French armies had been large,—either to procure subsistence more easily, or because the generals thought it better to put all the divisions in line, leaving it to their commanders to arrange them for battle. The reserves were small detachments, incapable of redeeming the day even if the enemy succeeded in overwhelming but a single division. Such was the state of affairs when Napoleon made his *debut* in Italy. His activity from the beginning worsted the Austrians and Piedmontese: free from useless incumbrances, his troops surpassed in mobility all modern armies. He conquered the Italian peninsula by a series of marches and strategic combats. His march on Vienna in 1797 was rash, but justified by the necessity of overcoming the Archduke Charles before he could receive reinforcements from the Rhine.

The campaign of 1800, still more characteristic of the man, marked a new era in the conception of plans of campaign and lines of operations. He adopted bold objective points, which looked to nothing less than the capture or destruction of whole armies. The orders of battle were less extended, and the more rational organization of armies in large bodies of two or three divisions was adopted. The system of modern strategy was here fully developed, and the campaigns of 1805 and 1806 were merely corollaries to the great problem solved in 1800. Tactically, the system of columns and skirmishers was too well adapted to the features of Italy not to meet with his approval.

It may now be a question whether the system of Napoleon is adapted to all capacities, epochs, and armies, or whether, on the contrary, there can be any return, in the light of the events of 1800 and 1809, to the old system of wars of position. After a comparison of the marches and camps of the Seven Years' War with those of the *seven weeks'* war,—as Napoleon called the campaign of 1806,—or with those of the three months which elapsed from the departure of the army from Boulogne in 1805 till its arrival in the plains of Moravia, the reader may easily decide as to the relative merits of the two systems.

The system of Napoleon was *to march twenty-five miles a day, to fight, and then to camp in quiet.* He told me that he knew no other method of conducting a war than this.

It may be said that the adventurous character of this great man, his personal situation, and the tone of the French mind, all concurred in urging him to undertakings which no other person, whether born upon a throne, or a general under the orders of his government, would ever dare to adopt. This is probably true; but between the extremes of very distant invasions, and wars of position, there is a proper mean, and, without imitating his impetuous audacity, we may pursue the line he has marked out. It is probable that the old system of wars of positions will for a long time be proscribed, or that, if adopted, it will be much modified and improved.

If the art of war is enlarged by the adoption of the system of marches, humanity, on the contrary, loses by it; for these rapid incursions and bivouacs of considerable masses, feeding upon the regions they overrun, are not materially different from the devastations of the barbarian hordes between the fourth and thirteenth centuries. Still, it is not likely that the system will be speedily renounced; for a great truth has been demonstrated by Napoleon's wars,—viz.: that remoteness is not a certain safeguard against invasion,—that a state to be secure must have a good system of fortresses and lines of defense, of reserves and military institutions, and, finally, a good system of government. Then the people may everywhere be organized as militia, and may serve as reserves to the active armies, which will render the latter more formidable; and the greater the strength of the armies the more necessary is the system of rapid operations and prompt results.

If, in time, social order assumes a calmer state,—if nations, instead of fighting for their existence, fight only for their interests, to acquire a natural frontier or to maintain the political equilibrium,—then a new right of nations may be agreed upon, and perhaps it will be possible to have armies on a less extensive scale. Then also we may see armies of from eighty to one hundred thousand men return to a mixed system of war,—a mean between the rapid incursions of Napoleon and the slow system of positions of the last century. Until then we must expect to retain this system of marches, which has produced so great results; for the first to renounce it in the presence of an active and capable enemy would probably be a victim to his indiscretion.

The science of marches now includes more than details, like the following, viz.: the order of the different arms in column, the time of departure and arrival, the precautions to be observed in the march, and the means of communication between the columns, all of which is a part of the duties of the staff of an army. Outside and beyond these very important details, there is a science of marches in the great operations of strategy. For instance, the march of Napoleon by the Saint-Bernard to fall upon the communications of Melas, those made in 1805

Chapter III. Strategy: Definition of Strategy
and the Fundamental Principle of War

by Donauwerth to cut off Mack, and in 1806 by Gera to turn the Prussians, the march of Suwaroff from Turin to the Trebbia to meet Macdonald, that of the Russian army on Taroutin, then upon Krasnoi, were decisive operations, not because of their relation to Logistics, but on account of their strategic relations.

Indeed, these skillful marches are but applications of the great principle of throwing the mass of the forces upon the decisive point; and this point is to be determined from the considerations given in Article XIX. What was the passage of the Saint-Bernard but a line of operations directed against an extremity of the strategic front of the enemy, and thence upon his line of retreat? The marches of Ulm and Jena were the same maneuvers; and what was Bluecher's march at Waterloo but an application of interior strategic lines?

From this it may be concluded that all strategic movements which tend to throw the mass of the army successively upon the different points of the front of operations of the enemy, will be skillful, as they apply the principle of overwhelming a smaller force by a superior one. The operations of the French in 1793 from Dunkirk to Landau, and those of Napoleon in 1796, 1809, and 1814, are models of this kind.

One of the most essential points in the science of modern marches, is to so combine the movements of the columns as to cover the greatest strategic front, when beyond the reach of the enemy, for the triple object of deceiving him as to the objective in view, of moving with ease and rapidity, and of procuring supplies with more facility. However, it is necessary in this case to have previously arranged the means of concentration of the columns in order to inflict a decisive blow.

This alternate application of extended and concentric movements is the true test of a great general.

There is another kind of marches, designated as *flank marches*, which deserves notice. They have always been held up as very dangerous; but nothing satisfactory has ever been written about them. If by the term *flank marches* are understood tactical maneuvers made upon the field of battle in view of the enemy, it is certain that they are very delicate operations, though sometimes successful; but if reference is made to ordinary strategic marches, I see nothing particularly dangerous in them, unless the most common precautions of Logistics be neglected. In a strategic movement, the two hostile armies ought to be separated by about two marches, (counting the distance which separates the advanced guards from the enemy and from their own columns.) In such a case there could be no danger in a strategic march from one point to another.

There are, however, two cases where such a march would be altogether inadmissible: the first is where the system of the line of operations, of the strategic lines, and of the front of operations is so chosen as to present the flank to the enemy during a whole operation. This was the famous project of marching upon Leipsic, leaving Napoleon and Dresden on the flank, which would, if carried out, have proved fatal to the allies. It was modified by the Emperor Alexander upon the solicitations of the author.

The second case is where the line of operations is very long, (as was the case with Napoleon at Borodino,) and particularly if this line affords but a single suitable route for retreat: then every flank movement exposing this line would be a great fault.

In countries abounding in secondary communications, flank movements are still less dangerous, since, if repulsed, safety may be found in a change of the line of operations. The physical and moral condition of the troops and the more or less energetic characters of the commanders will, of course, be elements in the determination of such movements.

The often-quoted marches of Jena and Ulm were actual flank maneuvers; so was that upon Milan after the passage of the Chiusella, and that of Marshal Paskevitch to cross the Vistula at Ossiek; and their successful issue is well known.

A tactical maneuver by the flank in the presence of the enemy is quite a different affair. Ney suffered for a movement of this kind at Dennewitz, and so did Marmont at Salamanca and Frederick at Kolin.

Nevertheless, the celebrated maneuver of Frederick at Leuthen was a true flank movement, but it was covered by a mass of cavalry concealed by the heights, and applied against an army which lay motionless in its camp; and it was so successful because at the time of the decisive shock Daun was taken in flank, and not Frederick.

In the old system of marching in column at platoon distance, where line of battle could be formed to the right or left without deployment, (by a right or left into line,) movements parallel to the enemy's line were not *flank marches*, because the flank of the column was the real front of the line of battle.

The famous march of Eugene within view of the French army, to turn the lines of Turin, was still more extraordinary than that of Leuthen, and no less successful.

In these different battles, the maneuvers were tactical and not strategic. The march of Eugene from Mantua to Turin was one of the greatest strategic operations of the age; but the case above referred to was a movement made to turn the French camp the evening before the battle.

ARTICLE XXV

DEPOTS OF SUPPLIES, AND THEIR RELATION TO MARCHES

The subject most nearly connected with the system of marches is the commissariat, for to march quickly and for a long distance food must be supplied; and the problem of supporting a numerous army in an enemy's

country is a very difficult one. It is proposed to discuss the relation between
the commissariat and strategy.

It will always be difficult to imagine how Darius and Xerxes subsisted their
immense armies in Thrace, where now it would be a hard task to supply thirty
thousand men. During the Middle Ages, the Greeks, barbarians, and more
lately the Crusaders, maintained considerable bodies of men in that country.
Caesar said that war should support war, and he is generally believed to have
lived at the expense of the countries he overran.

The Middle Ages were remarkable for the great migrations of all kinds, and
it would be interesting to know the numbers of the Huns, Vandals, Goths,
and Mongols who successively traversed Europe, and how they lived during
their marches. The commissariat arrangements of the Crusaders would also be
an interesting subject of research.

In the early periods of modern history, it is probable that the armies of
Francis I., in crossing the Alps into Italy, did not carry with them large stores
of provisions; for armies of their magnitude, of forty or fifty thousand men,
could easily find provisions in the rich valleys of the Ticino and Po.

Under Louis XIV. and Frederick II. the armies were larger; they fought on
their own frontiers, and lived from their storehouses, which were established
as they moved. This interfered greatly with operations, restricting the troops
within a distance from the depots dependent upon the means of transportation,
the rations they could carry, and the number of days necessary for wagons to
go to the depots and return to camp.

During the Revolution, depots of supply were abandoned from necessity.
The large armies which invaded Belgium and Germany lived sometimes in the
houses of the people, sometimes by requisitions laid upon the country, and often
by plunder and pillage. To subsist an army on the granaries of Belgium, Italy,
Swabia, and the rich banks of the Rhine and Danube, is easy,—particularly if
it marches in a number of columns and does not exceed one hundred or one
hundred and twenty thousand men; but this would be very difficult in some
other countries, and quite impossible in Russia, Sweden, Poland, and Turkey.
It may readily be conceived how great may be the rapidity and impetuosity of
an army where every thing depends only on the strength of the soldiers' legs.
This system gave Napoleon great advantages; but he abused it by applying it on
too large a scale and to countries where it was impracticable.

A general should be capable of making all the resources of the invaded
country contribute to the success of his enterprises: he should use the local
authorities, if they remain, to regulate the assessments so as to make them
uniform and legal, while he himself should see to their fulfillment. If the
authorities do not remain, he should create provisional ones of the leading men,
and endow them with extraordinary powers. The provisions thus acquired
should be collected at the points most convenient for the operations of the
army. In order to husband them, the troops may be quartered in the towns
and villages, taking care to reimburse the inhabitants for the extra charge thus

laid upon them. The inhabitants should also be required to furnish wagons to convey the supplies to the points occupied by the troops.

It is impossible to designate precisely what it will be prudent to undertake without having previously established these depots, as much depends upon the season, country, strength of the armies, and spirit of the people; but the following may be considered as general maxims:–

1. That in fertile and populous regions not hostile, an army of one hundred to one hundred and twenty thousand men, when so far distant from the enemy as to be able safely to recover a considerable extent of country, may draw its resources from it, during the time occupied by any single operation.

As the first operation never requires more than a month, during which time the great body of the troops will be in motion, it will be sufficient to provide, by depots of provisions, for the eventual wants of the army, and particularly for those of the troops obliged to remain at a particular point. Thus, the army of Napoleon, while half of it was besieging Ulm, would need bread until the surrender of the city; and if there had been a scarcity the operation might have failed.

2. During this time every effort should be made to collect the supplies obtained in the country, and to form depots, in order to subserve the wants of the army after the success of the operation, whether it take a position to recruit or whether it undertake a new enterprise.

3. The depots formed either by purchase or forced requisitions should be echeloned as much as possible upon three different lines of communication, in order to supply with more facility the wings of the army, and to extend as much as possible the area from which successive supplies are to be drawn, and, lastly, in order that the depots should be as well covered as possible. To this end, it would be well to have the depots on lines converging toward the principal line of operations, which will be generally found in the center. This arrangement has two real advantages: first, the depots are less exposed to the attempts of the enemy, as his distance from them is thereby increased; secondly, it facilitates the movements of the army in concentrating upon a single point of the line of operations to the rear, with a view of retaking the initiative from the enemy, who may have temporarily assumed the offensive and gained some advantage.

4. In thinly-settled and unproductive regions the army will lack its most necessary supplies: it will be prudent, in this case, not to advance too far from its depots, and to carry with it sufficient provisions to enable it, if compelled to do so, to fall back upon its lines of depots.

5. In national wars where the inhabitants fly and destroy every thing in their path, as was the case in Spain, Portugal, Russia, and Turkey, it is impossible to advance unless attended by trains of provisions and without having a sure base of supply near the front of operations. Under these circumstances a war of invasion becomes very difficult, if not impossible.

6. It is not only necessary to collect large quantities of supplies, but it is indispensable to have the means of conveying them with or after the army; and this is the greatest difficulty, particularly on rapid expeditions. To facilitate their transportation, the rations should consist of the most portable articles,—as biscuit, rice, &c.: the wagons should be both light and strong, so as to pass over all kinds of roads. It will be necessary to collect all the vehicles of the country, and to insure good treatment to their owners or drivers; and these vehicles should be arranged in parks at different points, so as not to take the drivers too far from their homes and in order to husband the successive resources. Lastly, the soldier must he habituated to carry with him several days' rations of bread, rice, or even of flour.

7. The vicinity of the sea is invaluable for the transportation of supplies; and the party which is master on this element can supply himself at will. This advantage, however, is not absolute in the case of a large continental army; for, in the desire to maintain communications with its depots, it may be drawn into operations on the coast, thus exposing itself to the greatest risks if the enemy maneuver with the mass of his forces upon the extremity opposite the sea. If the army advance too far from the coast, there will be danger of its communications being intercepted; and this danger increases with the progress of the army.

8. A continental army using the sea for transportation should base itself on the land, and have a reserve of provisions independent of its ships, and a line of retreat prepared on the extremity of its strategic front opposed to the sea.

9. Navigable streams and canals, when parallel to the line of operations of the army, render the transportation of supplies much easier, and also free the roads from the incumbrances of the numerous vehicles otherwise necessary. For this reason, lines of operations thus situated are the most favorable. The water-communications themselves are not in this case the lines of operations, as has been asserted: on the contrary, it is essential that the troops should be able to move at some distance from the river, in order to prevent the enemy from throwing back the exterior flank upon the river,—which might be as dangerous as if it were the sea.

In the enemy's country the rivers can scarcely ever be used for transportation, since the boats will probably be destroyed, and since a small body of men may easily embarrass the navigation. To render it sure, it is necessary to occupy both banks,—which is hazardous, as Mortier experienced at Dirnstein. In a friendly country the advantages of rivers are more substantial.

10. In default of bread or biscuit, the pressing wants of an army may be fed by cattle on the hoof; and these can generally be found, in populous countries, in numbers to last for some little time. This source of supply will, however, be soon exhausted; and, in addition, this plan leads to plunder. The requisitions for cattle should be well regulated; and the best plan of all is to supply the army with cattle purchased elsewhere.

I will end this article by recording a remark of Napoleon which may appear whimsical, but which is still not without reason. He said that in his first campaigns the enemy was so well provided that when his troops were in want of supplies he had only to fall upon the rear of the enemy to procure every thing in abundance. This is a remark upon which it would be absurd to found a system, but which perhaps explains the success of many a rash enterprise, and proves how much actual war differs from narrow theory.

ARTICLE XXVI

**THE DEFENSE OF FRONTIERS BY FORTS AND INTRENCHED LINES.—
WARS OF SIEGES**

Forts serve two principal purposes: first, to cover the frontiers; secondly, to aid the operations of the campaign.

The defense of frontiers is a problem generally somewhat indeterminate. It is not so for those countries whose borders are covered with great natural obstacles, and which present but few accessible points, and these admitting of defense by the art of the engineer. The problem here is simple; but in open countries it is more difficult. The Alps and the Pyrenees, and the lesser ranges of the Crapacks, of Riesengebirge, of Erzgebirge, of the Boehmerwald, of the Black Forest, of the Vosges, and of the Jura, are not so formidable that they cannot be made more so by a good system of fortresses.

Of all these frontiers, that separating France and Piedmont was best covered. The valleys of the Stura and Suza, the passes of Argentine, of Mont-Genevre, and of Mont-Cenis,—the only ones considered practicable,—were covered by masonry forts; and, in addition, works of considerable magnitude guarded the issues of the valleys in the plains of Piedmont. It was certainly no easy matter to surmount these difficulties.

These excellent artificial defenses will not always prevent the passage of an army, because the small works which are found in the gorges may be carried, or the enemy, if he be bold, may find a passage over some other route hitherto deemed impracticable. The passage of the Alps by Francis I.,—which is so well described by Gaillard,—Napoleon's passage of the Saint-Bernard, and the Splugen expedition, prove that there is truth in the remark of Napoleon, *that an army can pass wherever a titan can set his foot,*—a maxim not strictly true, but characteristic of the man, and applied by him with great success.

Other countries are covered by large rivers, either as a first line or as a second. It is, however, remarkable that such lines, apparently so well calculated to separate nations without interfering with trade and communication, are generally not part of the real frontier. It cannot be said that the Danube divides Bessarabia from the Ottoman empire as long as the Turks have a foothold in Moldavia. The Rhine was never the real frontier of France and Germany;

for the French for long periods held points upon the right bank, while the Germans were in possession of Mayence, Luxembourg, and the *tetes de ponts* of Manheim and Wesel on the left bank.

If, however, the Danube, the Rhine, Rhone, Elbe, Oder, Vistula, Po, and Adige be not exterior lines of the frontier, there is no reason why they should not be fortified as lines of permanent defense, wherever they permit the use of a system suitable for covering a front of operations.

An example of this kind is the Inn, which separates Bavaria from Austria: flanked on the south by the Tyrolese Alps, on the north by Bohemia and the Danube, its narrow front is covered by the three fortified places of Passau, Braunau, and Salzburg. Lloyd, with some poetic license, compares this frontier to two impregnable bastions whose curtain is formed of three fine forts and whose ditch is one of the most rapid of rivers. He has exaggerated these advantages; for his epithet of "impregnable" was decidedly disproved by the bloody events of 1800, 1805, and 1809.

The majority of the European states have frontiers by no means so formidable as that of the Alps and the Inn, being generally open, or consisting of mountains with practicable passes at a considerable number of points. We propose to give a set of general maxims equally applicable to all cases.

When the topography of a frontier is open, there should be no attempt to make a complete line of defense by building too many fortresses, requiring armies to garrison them, and which, after all, might not prevent an enemy from penetrating the country. It is much wiser to build fewer works, and to have them properly located, not with the expectation of absolutely preventing the ingress of the enemy, but to multiply the impediments to his progress, and, at the same time, to support the movements of the army which is to repel him.

If it be rare that a fortified place of itself absolutely prevents the progress of an army, it is, nevertheless, an embarrassment, and compels the army to detach a part of its force or to make *detours* in its march; while, on the other hand, it imparts corresponding advantages to the army which holds it, covers his depots, flanks, and movements, and, finally, is a place of refuge in case of need.

Fortresses thus exercise a manifest influence over military operations; and we now propose to examine their relations to strategy.

The first point to be considered is their location; the second lies in the distinction between the cases where an army can afford to pass the forts without a siege, and those where it will be necessary to besiege; the third point is in reference to the relations of an army to a siege which it proposes to cover.

As fortresses properly located favor military operations, in the same degree those which are unfortunately placed are disadvantageous. They are an incubus upon the army which is compelled to garrison them and the state whose men and money are wasted upon them. There are many in Europe in this category. It is bad policy to cover a frontier with fortresses very close together. This system has been wrongly imputed to Vauban, who, on the contrary, had a controversy with Louvois about the great number of points the latter desired to fortify. The maxims on this point are as follow:–

1. The fortified places should be in echelon, on three lines, and should extend from the frontiers toward the capital.[1] There should be three in the first line, as many in the second, and a large place in the third, near the center of the state. If there be four fronts, this would require, for a complete system, from twenty-four to thirty places.

It will be objected that this number is large, and that even Austria has not so many. It must be recollected that France has more than forty upon only a third of its frontiers, (from Besancon to Dunkirk,) and still has not enough on the third line in the center of the country. A Board convened for the purpose of considering the system of fortresses has decided quite recently that more were required. This does not prove that there were not already too many, but that certain points in addition should be fortified, while those on the first line, although too much crowded, may be maintained since they are already in existence. Admitting that France has two fronts from Dunkirk to Basel, one from Basel to Savoy, one from Savoy to Nice, in addition to the totally distinct line of the Pyrenees and the coast-line, there are six fronts, requiring forty to fifty places. Every military man will admit that this is enough, since the Swiss and coast fronts require fewer than the northeast. The system of arrangement of these fortresses is an important element of their usefulness. Austria has a less number, because she is bordered by the small German states, which, instead of being hostile, place their own forts at her disposal. Moreover, the number above given is what was considered necessary for a state having four fronts of nearly equal development. Prussia, being long and narrow, and extending from Koenigsberg almost to the gates of Metz, should not be fortified upon the same system as France, Spain, or Austria. Thus the geographical position and extent of states may either diminish or increase the number of fortresses, particularly when maritime forts are to be included.

2. Fortresses should always occupy the important strategic points already designated in Article XIX. As to their tactical qualities, their sites should not be commanded, and egress from them should be easy, in order to increase the difficulty of blockading them.

3. Those which possess the greatest advantages, either as to their own defense or for seconding the operations of an army, are certainly those situated on great rivers and commanding both banks. Mayence, Coblentz, and Strasbourg, including Kehl, are true illustrations and models of this kind. Places situated at the confluence of two great rivers command three different fronts, and hence are of increased importance. Take, for instance, Modlin. Mayence, when it

---

[1] The memorable campaign of 1829 is evidence of the value of such a system. If the Porte had possessed masonry forts in the defiles of the Balkan and a good fortress toward Faki, the Russians would not have reached Adrianople, and the affair would not have been so simple.

had on the left bank of the Main the fort of Gustavusburg, and Cassel on the right, was the most formidable place in Europe, but it required a garrison of twenty-five thousand men: so that works of this extent must be few in number.

4. Large forts, when encompassing populous and commercial cities, are preferable to small ones,—particularly when the assistance of the citizens can be relied on for their defense. Metz arrested the whole power of Charles V, and Lille for a whole year delayed Eugene and Marlborough. Strasbourg has many times proved the security of French armies. During the last wars these places were passed without being besieged by the invading forces, because all Europe was in arms against France; but one hundred and fifty thousand Germans having in their front one hundred thousand French could not penetrate to the Seine with impunity, leaving behind them these well-fortified points.

5. Formerly the operations of war were directed against towns, camps, and positions; recently they have been directed only against organized armies, leaving out of consideration all natural or artificial obstacles. The exclusive use of either of these systems is faulty: the true course is a mean between these extremes. Doubtless, it will always be of the first importance to destroy and disorganize all the armies of the enemy in the field, and to attain this end it may be allowable to pass the fortresses; but if the success be only partial it will be unwise to push the invasion too far. Here, also, very much depends upon the situation and respective strength of the armies and the spirit of the nations.

If Austria were the sole antagonist of France, she could not follow in the footsteps of the allies in 1814; neither is it probable that fifty thousand French will very soon risk themselves beyond the Noric Alps, in the very heart of Austria, as Napoleon did in 1797.[1] Such events only occur under exceptional circumstances.

6. It may be concluded from what precedes,—1st, that, while fortified places are essential supports, abuse in their application may, by dividing an army, weaken it instead of adding to its efficiency; 2d, that an army may, with the view of destroying the enemy, pass the line of these forts,—always, however, leaving a force to observe them; 3d, that an army cannot pass a large river, like the Danube or the Rhine, without reducing at least one of the fortresses on the river, in order to secure a good line of retreat. Once master of this place, the army may advance on the offensive, leaving detachments to besiege other places; and the chances of the reduction of those places increase as the army advances, since the enemy's opportunities of hindering the siege are correspondingly diminished.

7. While large places are much the most advantageous among a friendly people, smaller works are not without importance, not to arrest an enemy, who

---

[1]   Still, Napoleon was right in taking the offensive in the Frioul, since the Austrians were expecting a reinforcement from the Rhine of twenty thousand men, and of course it was highly important to beat the Archduke Charles before this force joined him. In view of the circumstances of the case, Napoleon's conduct was in accordance with the principles of war.

might mask them, but as they may materially aid the operations of an army in the field. The fort of Koenigstein in 1813 was as useful to the French as the fortress of Dresden, because it procured a *tete de pont* on the Elbe.

In a mountainous country, small, well-located forts are equal in value to fortified places, because their province is to close the passes, and not to afford refuge to armies: the little fort of Bard, in the valley of Aosta, almost arrested Napoleon's army in 1800.

8. It follows that each frontier should have one or two large fortresses as places of refuge, besides secondary forts and small posts to facilitate military operations. Walled cities with a shallow ditch may be very useful in the interior of a country, to contain depots, hospitals, &c, when they are strong enough to resist the attacks of any small bodies that may traverse the vicinity. They will be particularly serviceable if they can be defended by the militia, so as not to weaken the active army.

9. Large fortified places which are not in proper strategic positions are a positive misfortune for both the army and state.

10. Those on the sea-coast are of importance only in a maritime war, except for depots: they may even prove disastrous for a continental army, by holding out to it a delusive promise of support. Benningsen almost lost the Russian armies by basing them in 1807 on Koenigsberg,—which he did because it was convenient for supply. If the Russian army in 1812, instead of concentrating on Smolensk, had supported itself on Dunaburg and Riga, it would have been in danger of being forced into the sea and of being cut off from all its bases.

The relations between sieges and the operations of active armies are of two kinds. An invading army may pass by fortified places without attacking them, but it must leave a force to invest them, or at least to watch them; and when there are a number of them adjacent to each other it will be necessary to leave an entire corps d'armee, under a single commander, to invest or watch them as circumstances may require. When the invading army decides to attack a place, a sufficient force to carry on the siege will be assigned to this duty; the remainder may either continue its march or take a position to cover the siege.

Formerly the false system prevailed of encircling a city by a whole army, which buried itself in lines of circumvallation and contravallation. These lines cost as much in labor and expense as the siege itself. The famous case of the lines of Turin, which were fifteen miles in length, and, though guarded by seventy-eight thousand French, were forced by Prince Eugene with forty thousand men in 1706, is enough to condemn this ridiculous system.

Much as the recital of the immense labors of Caesar in the investment of Alise may excite our admiration, it is not probable that any general in our times will imitate his example. Nevertheless, it is very necessary for the investing force to strengthen its position by detached works commanding the routes by which the garrison might issue or by which the siege might be disturbed from without. This was done by Napoleon at Mantua, and by the Russians at Varna.

Experience has proved that the best way to cover a siege is to beat and pursue as far as possible the enemy's forces which could interfere. If the besieging force is numerically inferior, it should take up a strategic position covering all the avenues by which succor might arrive; and when it approaches, as much of the besieging force as can be spared should unite with the covering force to fall upon the approaching army and decide whether the siege shall continue or not.

Bonaparte in 1796, at Mantua, was a model of wisdom and skill for the operations of an army of observation.

INTRENCHED LINES

Besides the lines of circumvallation and contravallation referred to above, there is another kind, which is more extended than they are, and is in a measure allied to permanent fortifications, because it is intended to protect a part of the frontiers.

As a fortress or an intrenched camp may, as a temporary refuge for an army, be highly advantageous, so to the same degree is the system of intrenched lines absurd. I do not now refer to lines of small extent closing a narrow gorge, like Fussen and Scharnitz, for they may be regarded as forts; but I speak of extended lines many leagues in length and intended to wholly close a part of the frontiers. For instance, those of Wissembourg, which, covered by the Lauter flowing in front, supported by the Rhine on the right and the Vosges on the left, seemed to fulfill all the conditions of safety; and yet they were forced on every occasion when they were assailed.

The lines of Stollhofen, which on the right of the Rhine played the same part as those of Wissembourg on the left, were equally unfortunate; and those of the Queich and the Kinzig had the same fate.

The lines of Turin, (1706,) and those of Mayence, (1795,) although intended as lines of circumvallation, were analogous to the lines in question in their extent and in the fate which befell them. However well they may be supported by natural obstacles, their great extent paralyzes their defenders, and they are almost always susceptible of being turned. To bury an army in intrenchments, where it may be outflanked and surrounded, or forced in front even if secure from a flank attack, is manifest folly; and it is to be hoped that we shall never see another instance of it. Nevertheless, in our chapter on Tactics we will treat of their attack and defense.

It may be well to remark that, while it is absurd to use these extended lines, it would be equally foolish to neglect the advantages to be derived from detached works in increasing the strength of a besieging force, the safety of a position, or the defense of a defile.

THE CONNECTION OF INTRENCHED CAMPS AND TETES DE PONTS WITH STRATEGY

It would be out of place here to go into details as to the sites of ordinary camps and upon the means of covering them by advanced guards, or upon the advantages of field-fortifications in the defense of posts. Only fortified camps enter into the combinations of grand tactics, and even of strategy; and this they do by the temporary support they afford an army.

It may be seen by the example of the camp of Buntzelwitz, which saved Frederick in 1761, and by those of Kehl and Dusseldorf in 1796, that such a refuge may prove of the greatest importance. The camp of Ulm, in 1800, enabled Kray to arrest for a whole month the army of Moreau on the Danube; and Wellington derived great advantages from his camp of Torres-Vedras. The Turks were greatly assisted in defending the country between the Danube and the Balkan Mountains by the camp of Shumla.

The principal rule in this connection is that camps should be established on strategic points which should also possess tactical advantages. If the camp of Drissa was useless to the Russians in 1812, it was because it was not in a proper position in reference to their defensive system, which should have rested upon Smolensk and Moscow. Hence the Russians were compelled to abandon it after a few days.

The maxims which have been given for the determination of the great decisive strategic points will apply to all intrenched camps, because they ought only to be placed on such points. The influence of these camps is variable: they may answer equally well as points of departure for an offensive operation, as *tetes de ponts* to assure the crossing of a large river, as protection for winter quarters, or as a refuge for a defeated army.

However good may be the site of such a camp, it will always be difficult to locate it so that it may not be turned, unless, like the camp of Torres-Vedras, it be upon a peninsula backed by the sea. Whenever it can be passed either by the right or the left, the army will be compelled to abandon it or run the risk of being invested in it. The camp of Dresden was an important support to Napoleon for two months; but as soon as it was outflanked by the allies it had not the advantages even of an ordinary fortress; for its extent led to the sacrifice of two corps within a few days for want of provisions.

Despite all this, these camps, when only intended to afford temporary support to an army on the defensive, may still fulfill this end, even when the enemy passes by them, provided they cannot be taken in reverse,—that is, provided all their faces are equally safe from a *coup de main*. It is also important that they be established close to a fortress, where the depots may be safe, or which may cover the front of the camp nearest to the line of retreat.

In general terms, such a camp on a river, with a large *tete de pont* on the other side to command both banks, and near a large fortified city like Mayence or Strasbourg, is of undoubted advantage; but it will never be more than a temporary refuge, a means of gaining time and of collecting reinforcements. When the object is to drive away the enemy, it will be necessary to leave the camp and carry on operations in the open country.

The second maxim as to these camps is, that they are particularly advantageous to an army at home or near its base of operations. If a French army occupied an intrenched camp on the Elbe, it would be lost when the space between the Rhine and Elbe was held by the enemy; but if it were invested in an intrenched camp near Strasbourg, it might with a little assistance resume its superiority and take the field, while the enemy in the interior of France and between the relieving force and the intrenched army would have great difficulty in recrossing the Rhine.

We have heretofore considered these camps in a strategic light; but several German generals have maintained that they are suitable to cover places or to prevent sieges,—which appears to me to be a little sophistical. Doubtless, it will be more difficult to besiege a place when an army is encamped on its glacis; and it maybe said that the forts and camps are a mutual support; but, according to my view, the real and principal use of intrenched camps is always to afford, if necessary, a temporary refuge for an army, or the means of debouching offensively upon a decisive point or beyond a large river. To bury an army in such a camp, to expose it to the danger of being outflanked and cut off, simply to retard a siege, would be folly. The example of Wurmser, who prolonged the defense of Mantua, will be cited in opposition to this; but did not his army perish? And was this sacrifice really useful? I do not think so; for, the place having been once relieved and revictualed, and the siege-train having fallen into the hands of the Austrians, the siege was necessarily changed into a blockade, and the town could only be taken by reason of famine; and, this being the case, Wurmser's presence ought rather to have hastened than retarded its surrender.

The intrenched camp of the Austrians before Mayence in 1795 would, indeed, have prevented the siege of the place, if the French had possessed the means of carrying on a siege, as long as the Rhine had not been crossed; but as soon as Jourdan appeared on the Lahn, and Moreau in the Black Forest, it became necessary to abandon the camp and leave the place to its own means of defense. It would only be in the event of a fortress occupying a point such that it would be impossible for an army to pass it without taking it, that an intrenched camp, with the object of preventing an attack upon it, would be established; and what place in Europe is upon such a site?

So far from agreeing with these German authors, on the contrary, it seems to me that a very important question in the establishment of these camps near fortified places on a river, is whether they should be on the same bank as the place, or upon the other. When it is necessary to make a choice, by reason of the fact that the place cannot be located to cover both banks, I should decidedly prefer the latter.

To serve as a refuge or to favor a debouch, the camp should be on the bank of the river toward the enemy; and in this, case the principal danger to be feared is that the enemy might take the camp in reverse by passing the river at some other point; and if the fortress were upon the same bank us the camp, it would be of little service; while if upon the other bank, opposite to the camp, it would be almost impossible to take the latter in reverse. For instance, the Russians, who could not hold for twenty-four hours their camp of Drissa, would have defied the enemy for a long time if there had been a fortification on the right bank of the Dwina, covering the rear of the camp. So Moreau for three months, at Kehl, withstood all the efforts of the Archduke Charles; while if Strasbourg had not been there upon the opposite bank his camp would easily have been turned by a passage of the Rhine.

Indeed, it would be desirable to have the protection of the fortified place upon the other bank too; and a place holding both banks would fulfill this condition. The fortification of Coblentz, recently constructed, seems to introduce a new epoch. This system of the Prussians, combining the advantages of intrenched camps and permanent works, deserves attentive consideration; but, whatever may be its defects, it is nevertheless certain that it would afford immense advantages to an army intended to operate on the Rhine. Indeed, the inconvenience of intrenched camps on large rivers is that they are only very useful when beyond the river; and in this case they are exposed to the dangers arising from destruction of bridges (as happened to Napoleon at Essling,)—to say nothing of the danger of losing their provisions and munitions, or even of a front attack against which the works might not avail. The system of detached permanent works of Coblentz has the advantage of avoiding these dangers, by protecting the depots on the same bank as the army, and in guaranteeing to the army freedom from attack at least until the bridges be re-established. If the city were upon the right bank of the Rhine, and there were only an intrenched camp of field-works on the left bank, there would be no certainty of security either for the depots or the army. So, if Coblentz were a good ordinary fortress without detached forts, a large army could not so readily make it a place of refuge, nor would there be such facilities for debouching from it in the presence of an enemy. The fortress of Ehrenbreitstein, which is intended to protect Coblentz on the right bank, is so difficult of access that it would be quite easy to blockade it, and the egress of a force of any magnitude might be vigorously disputed.

Much has been recently said of a new system used by the Archduke Maximilian to fortify the intrenched camp of Linz,—by masonry towers. As

I only know of it by hearsay and the description by Captain Allard in the *Spectateur Militaire*, I cannot discuss it thoroughly. I only know that the system of towers used at Genoa by the skillful Colonel Andreis appeared to me to be useful, but still susceptible of improvements,—which the archduke seems to have added. We are told that the towers of Linz, situated in ditches and covered by the glacis, have the advantage of giving a concentrated horizontal fire and of being sheltered from the direct shot of the enemy. Such towers, if well flanked and connected by a parapet, may make a very advantageous camp,—always, however, with some of the inconveniences of closed lines. If the towers are isolated, and the intervals carefully covered by field-works, (to be thrown up when required,) they will make a camp preferable to one covered by ordinary redoubts, but not so advantageous as afforded by the large detached forts of Coblentz. These towers number thirty-two, eight of which are on the left bank, with a square fort commanding the Perlingsberg. Of these twenty-four on the right bank, some seven or eight are only half-towers. The circumference of this line is about twelve miles. The towers are between five hundred and six hundred yards apart, and will be connected, in case of war, by a palisaded covered way. They are of masonry, of three tiers of guns, with a barbette battery which is the principal defense, mounting eleven twenty-four pounders. Two howitzers are placed in the upper tier. Those towers are placed in a wide and deep ditch, the *deblais* of which forms a high glacis which protects the tower from direct shot; but I should think it would be difficult to protect the artillery from direct fire.

Some say that this has cost about three-fourths of what a complete bastioned enceinte, necessary to make Linz a fortress of the first rank, would have cost; others maintain that it has not cost more than a quarter as much as a bastioned work, and that it subserves, besides, an entirely different object. If these works are to resist a regular siege, they are certainly very defective; but, regarded as an intrenched camp to give refuge and an outlet upon both banks of the Danube for a large army, they are appropriate, and would be of great importance in a war like that of 1809, and, if existing then, would probably have saved the capital.

To complete a grand system, it would perhaps have been better to encircle Linz with a regular bastioned line, and then to have built seven or eight towers between the eastern salient and the mouth of the Traun, within a direct distance of about two and a half miles, so as to have included for the camp only the curved space between Linz, the Traun, and the Danube. Then the double advantage of a fortress of the first rank and a camp under its guns would have been united, and, even if not quite so large, would have answered for a large army, particularly if the eight towers on the left bank and the fort of Perlingsberg had been preserved.

TETES DE PONTS

tetes de ponts are the most important of all field-works. The difficulties of crossing a river, particularly a large one, in the face of the enemy, demonstrate abundantly the immense utility of such works, which can be less easily dispensed with than intrenched camps, since if the bridges are safe an army is insured from the disastrous events which may attend a rapid retreat across a large river.

tetes de ponts are doubly advantageous when they are as it were *keeps* for a large intrenched camp, and will be triply so if they also cover the bank opposite to the location of the camp, since then they will mutually support each other. It is needless to state that these works are particularly important in an enemy's country and upon all fronts where there are no permanent works. It may be observed that the principal difference between the system of intrenched camps and that of *tetes de ponts* is that the best intrenched camps are composed of detached and closed works, while *tetes de ponts* usually consist of contiguous works not closed. An intrenched line to admit of defense must be occupied in force throughout its whole extent, which would generally require a large army; if, on the contrary, the intrenchments are detached closed works, a comparatively small force can defend them.

The attack and defense of these works will be discussed in a subsequent part of this volume.

ARTICLE XXVIII
STRATEGIC OPERATIONS IN MOUNTAINS

A mountainous country presents itself, in the combinations of war, under four different aspects. It may be the whole theater of the war, or it may be but a zone; it may be mountainous throughout its whole extent, or there may be a line of mountains, upon emerging from which the army may debouch into large and rich plains.

If Switzerland, the Tyrol, the Noric provinces, some parts of Turkey and Hungary, Catalonia and Portugal, be excepted, in the European countries the mountains are in single ranges. In these cases there is but a difficult defile to cross,—a temporary obstacle, which, once overcome, is an advantage rather than an objection. In fact, the range once crossed and the war carried into the plains, the chain of mountains may be regarded as an eventual base, upon which the army may fall back and find a temporary refuge. The only essential precaution to be observed is, not to allow the enemy to anticipate the army on this line of retreat. The part of the Alps between France and Italy, and the Pyrenees, (which are not so high, though equally broad,) are of this nature. The mountains of Bohemia and of the Black Forest, and the Vosges, belong to this class. In Catalonia the mountains cover the whole country as far as the Ebro: if the war were limited to this province, the combinations would not be the same as if there were but a line of mountains. Hungary in this respect differs little from Lombardy and Castile; for if the Crapacks in the eastern and northern part are as marked a feature as the Pyrenees, they are still but a temporary obstacle, and an army overcoming it, whether debouching in the basin of the Waag, of the Neytra, or of the Theiss, or in the fields of Mongatsch, would have the vast plains between the Danube and the Theiss for a field of operations. The only difference would be in the roads, which in the Alps, though few in number, are excellent, while in Hungary there are none of much value. In its northern part, this chain, though not so high, becomes broader, and would seem to belong to that class of fields of operations which are wholly mountainous; but, as its evacuation may be compelled by decisive operations in the valleys of the Waag or the Theiss, it must be regarded as a temporary barrier. The attack and defense of this country, however, would be a strategic study of the most interesting character.

When an extremely mountainous country, such as the Tyrol or Switzerland, is but a zone of operations, the importance of these mountains is secondary, and they must be observed like a fortress, the armies deciding the great contests in the valleys. It will, of course, be otherwise if this be the whole field.

It has long been a question whether possession of the mountains gave control of the valleys, or whether possession of the valleys gave control of the mountains. The Archduke Charles, a very intelligent and competent judge, has declared for the latter, and has demonstrated that the valley of the Danube is the key of Southern Germany. However, in this kind of questions much depends upon the relative forces and their arrangement in the country. If sixty thousand French were advancing on Bavaria in presence of an equal force of Austrians, and the latter should throw thirty thousand men into the Tyrol, intending to replace them by reinforcements on its arrival on the Inn, it would be difficult for the French to push on as far as this line, leaving so large a force on its flanks masters of the outlets of Scharnitz, Fussen, Kufstein, and Lofers. But if the French force were one hundred and twenty thousand men, and had gained such successes as to establish its superiority over the army in its front, then it might leave a sufficient detachment to mask the passes of the Tyrol and extend its progress as far as Linz,—as Moreau did in 1800.

Thus far we have considered these mountainous districts as only accessory zones. If we regard them as the principal fields of operations, the strategic problem seems to be more complicated. The campaigns of 1799 and 1800 are equally rich in instruction on this branch of the art. In my account of them I have endeavored to bring out their teachings by a historical exposition of the events; and I cannot do better than refer my readers to it.

When we consider the results of the imprudent invasion of Switzerland by the French Directory, and its fatal influence in doubling the extent of the theater of operations and making it reach from the Texel to Naples, we cannot too much applaud the wisdom of France and Austria in the transactions which had for three centuries guaranteed the neutrality of Switzerland. Every one will be convinced of this by carefully studying the interesting campaigns of the Archduke Charles, Suwaroff, and Massena in 1799, and those of Napoleon and Moreau in 1800. The first is a model for operations upon an entirely mountainous field; the second is a model for wars in which the fate of mountainous countries is decided on the plains.

I will here state some of the deductions which seem to follow from this study.

When a country whose whole extent is mountainous is the principal theater of operations, the strategic combinations cannot be entirely based upon maxims applicable in an open country.

Transversal maneuvers to gain the extremity of the front of operations of the enemy here become always very difficult, and often impossible. In such a country a considerable army can be maneuvered only in a small number of valleys, where the enemy will take care to post advanced guards of sufficient strength to delay the army long enough to provide means for defeating the enterprise; and, as the ridges which separate these valleys will be generally crossed only by paths impracticable for the passage of an army, transversal marches can only be made by small bodies of light troops.

The important natural strategic points will be at the junction of the larger valleys or of the streams in those valleys, and will be few in number; and, if the defensive army occupy them with the mass of its forces, the invader will generally be compelled to resort to direct attacks to dislodge it.

However, if great strategic maneuvers in these cases be more rare and difficult, it by no means follows that they are less important. On the contrary, if the assailant succeed in gaining possession of one of these centers of communication between the large valleys upon the line of retreat of the enemy, it will be more serious for the latter than it would be in an open country; since the occupation of one or two difficult defiles will often be sufficient to cause the ruin of the whole army.

If the attacking party have difficulties to overcome, it must be admitted that the defense has quite as many, on account of the necessity of covering all the outlets by which an attack in force may be made upon the decisive points, and of the difficulties of the transversal marches which it would be compelled to make to cover the menaced points. In order to complete what I have said upon this kind of marches and the difficulties of directing them, I will refer to what Napoleon did in 1805 to cut off Mack from Ulm. If this operation was facilitated by the hundred roads which cross Swabia in all directions, and if it would have been impracticable in a mountainous country, for want of transversal routes, to make the long circuit from Donauwerth by Augsburg to Memmingen, it is also true that Mack could by these same hundred roads have effected his retreat with much greater facility than if he had been entrapped in one of the valleys of Switzerland or of the Tyrol, from which there was but a single outlet.

On the other hand, the general on the defensive may in a level country concentrate a large part of his forces; for, if the enemy scatter to occupy all the roads by which the defensive army may retire, it will be easy for the latter to crush these isolated bodies; but in a very mountainous country, where there are ordinarily but one or two principal routes into which other valleys open, even from the direction of the enemy, the concentration of forces becomes more difficult, since serious inconveniences may result if even one of these important valleys be not observed.

Nothing can better demonstrate the difficulty of strategic defense in mountainous regions than the perplexity in which we are involved when we attempt simply to give advice in such cases,—to say nothing of laying down maxims for them. If it were but a question of the defense of a single definite front of small extent, consisting of four or five converging valleys, the common junction of which is at a distance of two or three short marches from the summits of the ranges, it would be easier of solution. It would then be sufficient to recommend the construction of a good fort at the narrowest and least-easily turned point of each of these valleys. Protected by these forts, a few brigades of infantry should be stationed to dispute the passage, while half the army should be held in reserve at the junction, where it would be in position either to sustain the advanced guards most seriously threatened, or to fall upon the assailant with the whole force when he debouches. If to this be added good instructions to the commanders of the advanced guards, whether in assigning them the best point for rendezvous when their line of forts is pierced, or in directing them to continue to act in the mountains upon the flank of the enemy, the general on the defensive may regard himself as invincible, thanks to the many difficulties which the country offers to the assailant. But, if there be other fronts like this upon the right and left, all of which are to be defended, the problem is changed: the difficulties of the defense increase with the extent of the fronts, and this system of a cordon of forts becomes dangerous,—while it is not easy to adopt a better one.

We cannot be better convinced of these truths than by the consideration of the position of Massena in Switzerland in 1799. After Jourdan's defeat at Stockach, he occupied the line from Basel by Schaffhausen and Rheineck to Saint-Gothard, and thence by La Furca to Mont-Blanc. He had enemies in front of Basel, at Waldshut, at Schaffhausen, at Feldkirch, and at Chur; Bellegarde threatened the Saint-Gothard, and the Italian army menaced the Simplon and the Saint-Bernard. How was he to defend such a circumference? and how could he leave open one of these great valleys, thus risking every thing? From Rheinfelden to the Jura, toward Soleure, it was but two short marches, and there was the mouth of the trap in which the French army was placed. This was, then, the pivot of the defense. But how could he leave Schaffhausen unprotected? how abandon Rheineck and the Saint-Gothard?

how open the Valais and the approach by Berne, without surrendering the whole of Switzerland to the Coalition? And if he covered each point even by a brigade, where would be his army when he would need it to give battle to an approaching force? It is a natural system on a level theater to concentrate the masses of an army; but in the mountains such a course would surrender the keys of the country, and, besides, it is not easy to say where an inferior army could be concentrated without compromising it.

After the forced evacuation of the line of the Rhine and Zurich, it seemed that the only strategic point for Massena to defend was the line of the Jura. He was rash enough to stand upon the Albis,—a line shorter than that of the Rhine, it is true, but exposed for an immense distance to the attacks of the Austrians. If Bellegarde, instead of going into Lombardy by the Valtellina, had marched to Berne or made a junction with the archduke, Massena would have been ruined. These events seem to prove that if a country covered with high mountains be favorable for defense in a tactical point of view, it is different in a strategic sense, because it necessitates a division of the troops. This can only be remedied by giving them greater mobility and by passing often to the offensive.

General Clausewitz, whose logic is frequently defective, maintains, on the contrary, that, movements being the most difficult part in this kind of war, the defensive party should avoid them, since by such a course he might lose the advantages of the local defenses. He, however, ends by demonstrating that a passive defense must yield under an active attack,—which goes to show that the initiative is no less favorable in mountains than in plains. If there could be any doubt on this point, it ought to be dispelled by Massena's campaign in Switzerland, where he sustained himself only by attacking the enemy at every opportunity, even when he was obliged to seek him on the Grimsel and the Saint-Gothard. Napoleon's course was similar in 1796 in the Tyrol, when he was opposed to Wurmser and Alvinzi.

As for detailed strategic maneuvers, they may be comprehended by reading the events of Suwaroff's expedition by the Saint-Gothard upon the Muttenthal. While we must approve his maneuvers in endeavoring to capture Lecourbe in the valley of the Reuss, we must also admire the presence of mind, activity, and unyielding firmness which saved that general and his division. Afterward, in the Schachenthal and the Muttenthal, Suwaroff was placed in the same position as Lecourbe had been, and extricated himself with equal ability. Not less extraordinary was the ten days' campaign of General Molitor, who with four thousand men was surrounded in the canton of Glaris by more than thirty thousand allies, and yet succeeded in maintaining himself behind the Linth after four admirable fights. These events teach us the vanity of all theory *in details*, and also that in such a country a strong and heroic will is worth more than all the precepts in the world. After such lessons, need I say that one of the principal rules of this kind of war is, not to risk one's self in the valleys without securing the heights? Shall I say also that in this kind of war, more than in any other, operations should be directed upon the communications of the enemy? And, finally, that good temporary bases or lines of defense at the confluence of the great valleys, covered by strategic reserves, combined with great mobility and frequent offensive movements, will be the best means of defending the country?

I cannot terminate this article without remarking that mountainous countries are particularly favorable for defense when the war is a national one, in which the whole people rise up to defend their homes with the obstinacy which enthusiasm for a holy cause imparts: every advance is then dearly bought. But to be successful it is always necessary that the people be sustained by a disciplined force, more or less numerous: without this they must finally yield, like the heroes of Stanz and of the Tyrol.

The offensive against a mountainous country also presents a double case: it may either be directed upon a belt of mountains beyond which are extensive plains, or the whole theater may be mountainous.

In the first case there is little more to be done than this,—viz.: make demonstrations upon the whole line of the frontier, in order to lead the enemy to extend his defense, and then force a passage at the point which promises the greatest results. The problem in such a case is to break through a cordon which is strong less on account of the numbers of the defenders than from their position, and if broken at one point the whole line is forced. The history of Bard in 1800, and the capture of Leutasch and Scharnitz in 1805 by Ney, (who threw fourteen thousand men on Innspruck in the midst of thirty thousand Austrians, and by seizing this central point compelled them to retreat in all directions,) show that with brave infantry and bold commanders these famous mountain-ranges can generally be forced.

The history of the passage of the Alps, where Francis I. turned the army which was awaiting him at Suza by passing the steep mountains between Mont-Cenis and the valley of Queyras, is an example of those *insurmountable* obstacles which can always be surmounted. To oppose him it would have been necessary to adopt a system of cordon; and we have already seen what is to be expected of it. The position of the Swiss and Italians at Suza was even less wise than the cordon-system, because it inclosed them in a contracted valley without protecting the lateral issues. Their strategic plan ought to have been to throw troops into these valleys to defend the defiles, and to post the bulk of the army toward Turin or Carignano.

When we consider the *tactical* difficulties of this kind of war, and the immense advantages it affords the defense, we may be inclined to regard the concentration of a considerable force to penetrate by a single valley as an extremely rash maneuver, and to think that it ought to be divided into as many columns as there are practicable passes. In my opinion, this is one of the most dangerous of all illusions; and to confirm what I say it is only necessary to refer to the fate of the columns of Championnet at the battle of Fossano. If there be five or six roads on the menaced front, they should all, of course, be threatened; but the army should cross the chain in not more than two masses, and the routes which these follow should not be divergent; for if they were, the enemy might be able to defeat them separately. Napoleon's passage of the Saint-Bernard was wisely planned. He formed the bulk of his army on the center, with a division on each flank by Mont-Cenis and the Simplon, to divide the attention of the enemy and flank his march.

The invasion of a country entirely covered with mountains is a much greater and more difficult task than where a denouement may be accomplished by a decisive battle in the open country; for fields of battle for the deployment of large masses are rare in a mountainous region, and the war becomes a succession of partial combats. Here it would be imprudent, perhaps, to penetrate on a single point by a narrow and deep valley, whose outlets might be closed by the enemy and thus the invading army be endangered: it might penetrate by the wings on two or three lateral lines, whose outlets should not be too widely separated, the marches being so arranged that the masses may debouch at the junction of the valleys at nearly the same instant. The enemy should be driven from all the ridges which separate these valleys.

Of all mountainous countries, the tactical defense of Switzerland would be the easiest, if all her inhabitants were united in spirit; and with their assistance a disciplined force might hold its own against a triple number.

To give specific precepts for complications which vary infinitely with localities, the resources and the condition of the people and armies, would be absurd. History, well studied and understood, is the best school for this kind of warfare. The account of the campaign of 1799 by the Archduke Charles, that of the campaigns which I have given in my History of the Wars of the Revolution, the narrative of the campaign of the Grisons by Segur and Mathieu Dumas, that of Catalonia by Saint-Cyr and Suchet, the campaign of the Duke de Rohan in Valtellina, and the passage of the Alps by Gaillard, (Francis I.,) are good guides in this study.

ARTICLE XXIX
GRAND INVASIONS AND DISTANT EXPEDITIONS

There are several kinds of distant expeditions. The first are those which are merely auxiliary and belong to wars of intervention. The second are great continental invasions, through extensive tracts of country, which may be either friendly, neutral, doubtful, or hostile. The third are of the same nature, but made partly on land, partly by sea by means of numerous fleets. The fourth class comprises those beyond the seas, to found, defend, or attack distant colonies. The fifth includes the great descents, where the distance passed over is not very great, but where a powerful state is attacked.

As to the first, in a strategic point of view, a Russian army on the Rhine or in Italy, in alliance with the German States, would certainly be stronger and more favorably situated than if it had reached either of these points by passing over hostile or even neutral territory; for its base, lines of operations, and eventual points of support will be the same as those of its allies; it may find refuge behind their lines of defense, provisions in their depots, and munitions in their arsenals;—while in the other case its resources would be upon the Vistula or the Niemen, and it might afford another example of the sad fate of many of these great invasions.

In spite of the important difference between a war in which a state is merely an auxiliary, and a distant invasion undertaken for its own interest and with its own resources, there are, nevertheless, dangers in the way of these auxiliary armies, and perplexity for the commander of all the armies,—particularly if he belong to the state which is not a principal party; as may be learned from the campaign of 1805. General Koutousoff advanced on the Inn to the boundaries of Bavaria with thirty thousand Russians, to effect a junction with Mack, whose army in the mean time had been destroyed, with the exception of eighteen thousand men brought back from Donauwerth by Kienmayer. The Russian general thus found himself with fifty thousand men exposed to the impetuous activity of Napoleon with one hundred and fifty thousand, and, to complete his misfortune, he was separated from his own frontiers by a distance of about seven hundred and fifty miles. His position would have been hopeless if fifty thousand men had not arrived to reinforce him. The battle of Austerlitz—due to a fault of Weyrother—endangered the Russian army anew, since it was so far from its base. It almost became the victim of a distant alliance; and it was only peace that gave it the opportunity of regaining its own country.

The fate of Suwaroff after the victory of Novi, especially in the expedition to Switzerland, and that of Hermann's corps at Bergen in Holland, are examples which should be well studied by every commander under such circumstances. General Benningsen's position in 1807 was less disadvantageous, because, being between the Vistula and the Niemen, his communications with his base were preserved and his operations were in no respect dependent upon his allies. We may also refer to the fate of the French in Bohemia and Bavaria in 1742, when Frederick the Great abandoned them and made a separate peace. In this case the parties were allies rather than auxiliaries; but in the latter relation the political ties are never woven so closely as to remove all points of dissension which may compromise military operations. Examples of this kind have been cited in Article XIX., on political objective points.

History alone furnishes us instruction in reference to distant invasions across extensive territories. When half of Europe was covered with forests, pasturages, and flocks, and when only horses and iron were necessary to transplant whole nations from one end of the continent to the other, the Goths, Huns, Vandals, Normans, Arabs, and Tartars overran empires in succession. But since the invention of powder and artillery and the organization of formidable standing armies, and particularly since civilization and statesmanship have brought nations closer together and have taught them the necessity of reciprocally sustaining each other, no such events have taken place.

Besides these migrations of nations, there were other expeditions in the Middle Ages, which were of a more military character, as those of Charlemagne and others. Since the invention of powder there have been scarcely any, except the advance of Charles VIII. to Naples, and of Charles XII. into the Ukraine, which can be called distant invasions; for the campaigns of the Spaniards in Flanders and of the Swedes in Germany were of a particular kind. The first was a civil war, and the Swedes were only auxiliaries to the Protestants of Germany; and, besides, the forces concerned in both were not large. In modern times no one but Napoleon has dared to transport the armies of half of Europe from the Rhine to the Volga; and there is little danger that he will be imitated.

Apart from the modifications which result from great distances, all invasions, after the armies arrive upon the actual theater, present the same operations as all other wars. As the chief difficulty arises from these great distances, we should recall our maxims on deep lines of operations, strategic reserves, and eventual bases, as the only ones applicable; and here it is that their application is indispensable, although even that will not avert all danger. The campaign of 1812, although so ruinous to Napoleon, was a model for a distant invasion. His care in leaving Prince Schwarzenberg and Reynier on the Bug, while Macdonald, Oudinot, and Wrede guarded the Dwina, Victor covered Smolensk, and Augereau was between the Oder and Vistula, proves that he had neglected no humanly possible precaution in order to base himself safely; but it also proves that the greatest enterprises may fail simply on account of the magnitude of the preparations for their success.

If Napoleon erred in this contest, it was in neglecting diplomatic precautions; in not uniting under one commander the different bodies of troops on the Dwina and Dnieper; in remaining ten days too long at Wilna; in giving the command of his right to his brother, who was unequal to it; and in confiding to Prince Schwarzenberg a duty which that general could not perform with the devotedness of a Frenchman. I do not speak now of his error in remaining in Moscow after the conflagration, since then there was no remedy for the misfortune; although it would not have been so great if the retreat had taken place immediately. He has also been accused of having too much despised distances, difficulties, and men, in pushing on as far as the Kremlin. Before passing judgment upon him in this matter, however, we ought to know the real motives which induced him to pass Smolensk, instead of wintering there as he had intended, and whether it would have been possible for him to remain between that city and Vitebsk without having previously defeated the Russian army.

It is doubtless true that Napoleon neglected too much the resentment of Austria, Prussia, and Sweden, and counted too surely upon a *denouement* between Wilna and the Dwina. Although he fully appreciated the bravery of the Russian armies, he did not realize the spirit and energy of the people. Finally, and chiefly, instead of procuring the hearty and sincere concurrence of a military state, whose territories would have given him a sure base for his attack upon the colossal power of Russia, he founded his enterprise upon the co-operation of a brave and enthusiastic but fickle people, and besides, he neglected to turn to the greatest advantage this ephemeral enthusiasm.

The fate of all such enterprises makes it evident that the capital point for their success, and, in fact, the only maxim to be given, is "never to attempt them without having secured the hearty and constant alliance of a respectable power near enough the field of operations to afford a proper base, where supplies of every kind may be accumulated, and which may also in case of reverse serve as a refuge and afford new means of resuming the offensive." As to the precautions to be observed in these operations, the reader is referred to Articles XXI. and XXII., on the safety of deep lines of operations and the establishment of eventual bases, as giving all the military means of lessening the danger; to these should be added a just appreciation of distances, obstacles, seasons, and countries,—in short, accuracy in calculation and moderation in success, in order that the enterprise may not be carried too far. We are far from thinking that any purely military maxims can insure the success of remote invasions: in four thousand years only five or six have been successful, and in a hundred instances they have nearly ruined nations and armies.

Expeditions of the third class, partly on land, partly by sea, have been rare since the invention of artillery, the Crusades being the last in date of occurrence; and probably the cause is that the control of the sea, after having been held in succession by several secondary powers, has passed into the hands of England, an insular power, rich in ships, but without the land-forces necessary for such expeditions.

It is evident that from both of these causes the condition of things now is very different from that existing when Xerxes marched to the conquest of Greece, followed by four thousand vessels of all dimensions, or when Alexander marched from Macedonia over Asia Minor to Tyre, while his fleet coasted the shore.

Nevertheless, if we no longer see such invasions, it is very true that the assistance of a fleet of men-of-war and transports will always be of immense value to any army on shore when the two can act in concert. Still, sailing-ships are an uncertain resource, for their progress depends upon the winds,—which may be unfavorable: in addition, any kind of fleet is exposed to great dangers in storms, which are not of rare occurrence.

The more or less hostile tone of the people, the length of the line of operations, and the great distance of the principal objective point, are the only points which require any deviation from the ordinary operations of war.

Invasions of neighboring states, if less dangerous than distant ones, are still not without great danger of failure. A French army attacking Cadiz might find a tomb on the Guadalquivir, although well based upon the Pyrenees and possessing intermediate bases upon the Ebro and the Tagus. Likewise, the army which in 1809 besieged Komorn in the heart of Hungary might have been destroyed on the plains of Wagram without going as far as the Beresina. The antecedents, the number of disposable troops, the successes already gained, the state of the country, will all be elements in determining the extent of the enterprises to be undertaken; and to be able to proportion them well to his resources, in view of the attendant circumstances, is a great talent in a general. Although diplomacy does not play so important a part in these invasions as in those more distant, it is still of importance; since, as stated in Article VI., there is no enemy, however insignificant, whom it would not be useful to convert into an ally. The influence which the change of policy of the Duke of Savoy in 1706 exercised over the events of that day, and the effects of the stand taken by Maurice of Saxony in 1551, and of Bavaria in 1813, prove clearly the importance of securing the strict neutrality of all states adjoining the theater of war, when their co-operation cannot be obtained.

EPITOME OF STRATEGY

The task which I undertook seems to me to have been passably fulfilled by what has been stated in reference to the strategic combinations which enter ordinarily into a plan of campaign. We have seen, from the definition at the beginning of this chapter, that, in the most important operations in war, *strategy* fixes the direction of movements, and that we depend upon *tactics* for their execution. Therefore, before treating of these mixed operations, it will be well to give here the combinations of grand tactics and of battles, as well as the maxims by the aid of which the application of the fundamental principle of war may be made.

By this method these operations, half strategic and half tactical, will be better comprehended as a whole; but, in the first place, I will give a synopsis of the contents of the preceding chapter.

From the different articles which compose it, we may conclude that the manner of applying the general principle of war to all possible theaters of operations is found in what follows:–

1. In knowing how to make the best use of the advantages which the reciprocal directions of the two bases of operations may afford, in accordance with Article XVIII.

2. In choosing, from the three zones ordinarily found in the strategic field, that one upon which the greatest injury can be done to the enemy with the least risk to one's self.

3. In establishing well, and giving a good direction to, the lines of operations; adopting for defense the concentric system of the Archduke Charles in 1796 and of Napoleon in 1814; or that of Soult in 1814, for retreats parallel to the frontiers.

On the offensive we should follow the system which led to the success of Napoleon in 1800, 1805, and 1806, when he directed his line upon the extremity of the strategic front; or we might adopt his plan which was successful in 1796, 1809, and 1814, of directing the line of operations upon the center of the strategic front: all of which is to be determined by the respective positions of the armies, and according to the maxims presented in Article XXI.

4. In selecting judicious eventual lines of maneuver, by giving them such directions as always to be able to act with the greater mass of the forces, and to prevent the parts of the enemy from concentrating or from affording each other mutual support.

5. In combining, in the same spirit of centralization, all strategic positions, and all large detachments made to cover the most important strategic points of the theater of war.

6. In imparting to the troops the greatest possible mobility and activity, so as, by their successive employment upon points where it may be important to act, to bring superior force to bear upon fractions of the hostile army.

The system of rapid and continuous marches multiplies the effect of an army, and at the same time neutralizes a great part of that of the enemy's, and is often sufficient to insure success; but its effect will be quintupled if the marches be skillfully directed upon the decisive strategic points of the zone of operations, where the severest blows to the enemy can be given.

However, as a general may not always be prepared to adopt this decisive course to the exclusion of every other, he must then be content with attaining a part of the object of every enterprise, by rapid and successive employment of his forces upon isolated bodies of the enemy, thus insuring their defeat. A general who moves his masses rapidly and continually, and gives them proper directions, may be confident both of gaining victories and of securing great results therefrom.

The oft-cited operations of 1809 and 1814 prove these truths most satisfactorily, as also does that ordered by Carnot in 1793, already mentioned in Article XXIV., and the details of which may be found in Volume IV. of my History of the Wars of the Revolution. Forty battalions, carried successively from Dunkirk to Menin, Maubeuge, and Landau, by reinforcing the armies already at those points, gained four victories and saved France. The whole science of marches would have been found in this wise operation had it been directed upon the decisive strategic point. The Austrian was then the principal army of the Coalition, and its line of retreat was upon Cologne: hence it was upon the Meuse that a general effort of the French would have inflicted the most severe blow. The Committee of Public Safety provided for the most pressing danger, and the maneuver contains half of the strategic principle; the other half consists in giving to such efforts the most decisive direction, as Napoleon did at Ulm, at Jena, and at Ratisbon. The whole of strategy is contained in these four examples.

It is superfluous to add that one of the great ends of strategy is to be able to assure real advantages to the army by preparing the theater of war most favorable for its operations, if they take place in its own country, by the location of fortified places, of intrenched camps, and of *tetes de ponts*, and by the opening of communications in the great decisive directions: these constitute not the least interesting part of the science. We have already seen how we are to recognize these lines and these decisive points, whether permanent or temporary. Napoleon has afforded instruction on this point by the roads of the Simplon and Mont-Cenis; and Austria since 1815 has profited by it in the roads from the Tyrol to Lombardy, the Saint-Gothard, and the Splugen, as well as by different fortified places projected or completed.

COMMENTARY ON CHAPTER III

Chapter 3 is the heart of The Summary of the Art of War and the longest chapter in the book. This chapter is what established Jomini's reputation and what has received the most criticism. In the introduction to this chapter, Jomini provides what he regards as the fundamental principles involved in winning wars, repeated here:

THE FUNDAMENTAL PRINCIPLES OF WAR.

It is proposed to show that there is one great principle underlying all the operations of war—a principle which must be followed in all good combinations. It is embraced in the following maxims:

1. To throw by strategic movements the mass of an army, successively, upon the decisive points of a theater of war, and also upon the communications of the enemy as much as possible without compromising one's own.

2. To maneuver to engage fractions of the hostile army with the bulk of one's forces.

3. On the battle-field, to throw the mass of the forces upon the decisive point, or upon that portion of the hostile line which it is of the first importance to overthrow.

4. To so arrange that these masses shall not only be thrown upon the decisive point, but that they shall engage at the proper times and with energy (cite source).

Jomini then states that the rest of the chapter will reveal the proper combinations necessary to apply these principles. Most criticism of Jomini stems from a belief that he claims that a mechanical application of the maneuvers in this chapter will automatically lead to victory. To be fair to his critics, much of Jomini's rhetoric lends itself to this interpretation. In addition, critics have attacked the pedantic way in which Jomini defines the various strategic lines and points as well as bases and objectives in this chapter. The amount of time given to these matters in this section of the book has led to the accusation that Jomini was obsessed with geographical objectives and geometric movements, ignoring the fluidity inherent in a campaign against even a moderately active foe. There is some justice in these criticisms as well. Jomini's perfect campaign would look like Ulm, where Napoleon destroyed Mack by maneuver alone—a feat rarely achieved by even the greatest leaders. A careful reading of Chapter 3, however, reveals that Jomini's penchant for maneuver had the goal of gaining a concrete advantage over the enemy in battle, as the fundamental principles listed above demonstrate. This emphasis is reinforced in the post-script to this chapter, entitled "Epitome of Strategy," in which Jomini gives proper emphasis to the virtues of speed and decisiveness in offensive maneuvers, demonstrating that despite his geometrical diagrams, Jomini was not entirely an 18th century formalist.

CHAPTER IV

GRAND TACTICS AND BATTLES

*B*attles are the actual conflicts of armies contending about great questions of national policy and of strategy. Strategy directs armies to the decisive points of a zone of operations, and influences, in advance, the results of battles; but tactics, aided by courage, by genius and fortune, gains victories.

Grand tactics is the art of making good combinations preliminary to battles, as well as during their progress. The guiding principle in tactical combinations, as in those of strategy, is to bring the mass of the force in hand against a part of the opposing army, and upon that point the possession of which promises the most important results.

Battles have been stated by some writers to be the chief and deciding features of war. This assertion is not strictly true, as armies have been destroyed by strategic operations without the occurrence of pitched battles, by a succession of inconsiderable affairs. It is also true that a complete and decided victory may give rise to results of the same character when there may have been no grand strategic combinations.

The results of a battle generally depend upon a union of causes which are not always within the scope of the military art: the nature of the order of battle adopted, the greater or less wisdom displayed in the plan of the battle, as well as the manner of carrying out its details, the more or less loyal and enlightened co-operation of the officers subordinate to the commander-in-chief, the cause of the contest, the proportions and quality of the troops, their greater or less enthusiasm, superiority on the one side or the other in artillery or cavalry, and the manner of handling these arms; but it is the *morale* of armies, as well as of nations, more than any thing else, which makes victories and their results decisive. Clausewitz commits a grave error in asserting that a battle not characterized by a maneuver to turn the enemy cannot result in a complete victory. At the battle of Zama, Hannibal, in a few brief hours, saw the fruits of twenty years of glory and success vanish before his eyes, although Scipio never had a thought of turning his position. At Rivoli the turning-party was completely beaten; nor was the maneuver more successful at Stockach in 1799, or at Austerlitz in 1805. As is evident from Article XXXII., I by no means intend to discourage the use of that maneuver, being, on the contrary, a constant advocate of it; but it is very important to know how to use it skillfully and opportunely, and I am, moreover, of opinion that if it be a general's design to make himself master of his enemy's communications while at the same time holding his own, he would do better to employ strategic than tactical combinations to accomplish it.

There are three kinds of battles: 1st, defensive battles, or those fought by armies in favorable positions taken up to await the enemy's attack; 2d, offensive battles, where one army attacks another in position; 3d, battles fought unexpectedly, and resulting from the collision of two armies meeting on the march. We will examine in succession the different combinations they present.

ARTICLE XXX
POSITIONS AND DEFENSIVE BATTLES

When an army awaits an attack, it takes up a position and forms its line of battle. From the general definitions given at the beginning of this work, it will appear that I make a distinction between *lines of battle* and *orders of battle*,—things which have been constantly confounded. I will designate as a *line of battle* the position occupied by battalions, either deployed or in columns of attack, which an army will take up to hold a camp and a certain portion of ground where it will await attack, having no particular project in view for the future: it is the right name to give to a body of troops formed with proper tactical intervals and distances upon one or more lines, as will be more fully explained in Article XLIII. On the contrary, I will designate as an *order of battle* an arrangement of troops indicating an intention to execute a certain maneuver; as, for example, the parallel order, the oblique order, the perpendicular order.

This nomenclature, although new, seems necessary to keeping up a proper distinction between two things which should by no means be confounded.[1] From the nature of the two things, it is evident that the *line of battle* belongs especially to defensive arrangements; because an army awaiting an attack without knowing what or where it will be must necessarily form a rather indefinite and objectless line of battle. *order of battle*, on the contrary, indicating an arrangement of troops formed with an intention of fighting while executing some maneuver previously determined upon, belongs more particularly to

---

[1] It is from no desire to make innovations that I have modified old terms or made new. In the development of a science, it is wrong for the same word to designate two very different things; and, if we continue to apply the term *order of battle* to the disposition of troops in line, it must be improper to designate certain important maneuvers by the terms *oblique order of battle*, *concave order of battle*, and it becomes necessary to use instead the terms *oblique system of battle*, &c. I prefer the method of designation I have adopted. The *order of battle* on paper may take the name *plan of organization*, and the ordinary formation of troops upon the ground will then be called *line of battle*.

offensive dispositions. However, it is by no means pretended that the line of battle is exclusively a defensive arrangement; for a body of troops may in this formation very well proceed to the attack of a position, while an army on the defensive may use the oblique order or any other. I refer above only to ordinary cases.

Without adhering strictly to what is called the system of a war of positions, an army may often find it proper to await the enemy at a favorable point, strong by nature and selected beforehand for the purpose of there fighting a defensive battle. Such a position may be taken up when the object is to cover an important objective point, such as a capital, large depots, or a decisive strategic point which controls the surrounding country, or, finally, to cover a siege.

There are two kinds of positions,—the *strategic*, which has been discussed in Article XX., and the *tactical*. The latter, again, are subdivided. In the first place, there are intrenched positions occupied to await the enemy under cover of works more or less connected,—in a word, intrenched camps. Their relations to strategic operations have been treated in Article XXVII., and their attack and defense are discussed in Article XXXV. Secondly, we have positions naturally strong, where armies encamp for the purpose of gaining a few days' time. Third and last are open positions, chosen in advance to fight on the defensive. The characteristics to be sought in these positions vary according to the object in view: it is, however, a matter of importance not to be carried away by the mistaken idea, which prevails too extensively, of giving the preference to positions that are very steep and difficult of access,—quite suitable places, probably, for temporary camps, but not always the best for battle-grounds. A position of this kind, to be really strong, must be not only steep and difficult of access, but should be adapted to the end had in view in occupying it, should offer as many advantages as possible for the kind of troops forming the principal strength of the army, and, finally, the obstacles presented by its features should be more disadvantageous for the enemy than for the assailed. For example, it is certain that Massena, in taking the strong position of the Albis, would have made a great error if his chief strength had been in cavalry and artillery; whilst it was exactly what was wanted for his excellent infantry. For the same reason, Wellington, whose whole dependence was in the fire of his troops, made a good choice of position at Waterloo, where all the avenues of approach were well swept by his guns. The position of the Albis was, moreover, rather a strategic position, that of Waterloo being simply a battle-ground.

The rules to be generally observed in selecting tactical positions are the following:–

1. To have the communications to the front such as to make it easier to fall upon the enemy at a favorable moment than for him to approach the line of battle.

2. To give the artillery all its effect in the defense.

3. To have the ground suitable for concealing the movements of troops between the wings, that they may be massed upon any point deemed the proper one.

4. To be able to have a good view of the enemy's movements.

5. To have an unobstructed line of retreat.

6. To have the flanks well protected, either by natural or artificial obstacles, so as to render impossible an attack upon their extremities, and to oblige the enemy to attack the center, or at least some point of the front.

This is a difficult condition to fulfill; for, if an army rests on a river, or a mountain, or an impenetrable forest, and the smallest reverse happens to it, a great disaster may be the result of the broken line being forced back upon the very obstacles which seemed to afford perfect protection. This danger—about which there can be no doubt—gives rise to the thought that points admitting an easy defense are better on a battle field than insurmountable obstacles.[1]

7. Sometimes a want of proper support for the flanks is remedied by throwing a crotchet to the rear. This is dangerous; because a crotchet stuck on a line hinders its movements, and the enemy may cause great loss of life by placing his artillery in the angle of the two lines prolonged. A strong reserve in close column behind the wing to be guarded from assault seems better to fulfill the required condition than the crotchet; but the nature of the ground must always decide in the choice between the two methods. Full details on this point are given in the description of the battle of Prague, (Chapter II. of the Seven Years' War.)

8. We must endeavor in a defensive position not only to cover the flanks, but it often happens that there are obstacles on other points of the front, of such a character as to compel an attack upon the center. Such a position will always be one of the most advantageous for defense,—as was shown at Malplaquet and Waterloo. Great obstacles are not essential for this purpose, as the smallest accident of the ground is sometimes sufficient: thus, the insignificant rivulet of Papelotte forced Ney to attack Wellington's center, instead of the left as he had been ordered.

When a defense is made of such a position, care must be taken to hold ready for movement portions of the wings thus covered, in order that they may take part in the action instead of remaining idle spectators of it.

The fact cannot be concealed, however, that all these means are but palliatives; and the best thing for an army standing on the defensive is to *know* how to take the offensive at a proper time, and *to take it*. Among the conditions

---

1   The park of Hougoumont, the hamlet of La Haye Sainte, and the rivulet of Papelotte were for Ney more serious obstacles than the famous position of Elchingen, where he forced a passage of the Danube, in 1805, upon the ruins of a burnt bridge. It may perhaps be said that the courage of the defenders in the two cases was not the same; but, throwing out of consideration this chance, it must be granted that the difficulties of a position, when properly taken advantage of, need not be insurmountable in order to render the attack abortive. At Elchingen the great height and steepness of the banks, rendering the fire almost ineffectual, were more disadvantageous than useful in the defense.

to be satisfied by a defensive position has been mentioned that of enabling an easy and safe retreat; and this brings us to an examination of a question presented by the battle of Waterloo. Would an army with its rear resting upon a forest, and with a good road behind the center and each wing, have its retreat compromised, as Napoleon imagined, if it should lose the battle? My own opinion is that such a position would be more favorable for a retreat than an entirely open field; for a beaten army could not cross a plain without exposure to very great danger. Undoubtedly, if the retreat becomes a rout, a portion of the artillery left in battery in front of the forest would, in all probability, be lost; but the infantry and cavalry and a great part of the artillery could retire just as readily as across a plain. There is, indeed, no better cover for an orderly retreat than a forest,—this statement being made upon the supposition that there are at least two good roads behind the line, that proper measures for retreat have been taken before the enemy has had an opportunity to press too closely, and, finally, that the enemy is not permitted by a flank movement to be before the retreating army at the outlet of the forest, as was the case at Hohenlinden. The retreat would be the more secure if, as at Waterloo, the forest formed a concave line behind the center; for this re-entering would become a place of arms to receive the troops and give them time to pass off in succession on the main roads.

When discussing strategic operations, mention was made of the varying chances which the two systems, the *defensive* and the *offensive*, give rise to; and it was seen that especially in strategy the army taking the initiative has the great advantage of bringing up its troops and striking a blow where it may deem best, whilst the army which acts upon the defensive and awaits an attack is anticipated in every direction, is often taken unawares, and is always obliged to regulate its movements by those of the enemy. We have also seen that in tactics these advantages are not so marked, because in this case the operations occupy a smaller extent of ground, and the party taking the initiative cannot conceal his movements from the enemy, who, instantly observing, may at once counteract them by the aid of a good reserve. Moreover, the party advancing upon the enemy has against him all the disadvantages arising from accidents of ground that he must pass before reaching the hostile line; and, however flat a country it may be, there are always inequalities of the surface, such as small ravines, thickets, hedges, farm-houses, villages, &c., which must either be taken possession of or be passed by. To these natural obstacles may also be added the enemy's batteries to be carried, and the disorder which always prevails to a greater or less extent in a body of men exposed to a continued fire either of musketry or artillery. Viewing the matter in the light of these facts, all must agree that in tactical operations the advantages resulting from taking the initiative are balanced by the disadvantages.

However undoubted these truths may be, there is another, still more manifest, which has been demonstrated by the greatest events of history. Every army which maintains a strictly defensive attitude must, if attacked, be at last driven from its position; whilst by profiting by all the advantages of the defensive system, and holding itself ready to take the offensive when occasion offers, it may hope for the greatest success. A general who stands motionless to receive his enemy, keeping strictly on the defensive, may fight ever so bravely, but he must give way when properly attacked. It is not so, however, with a general who indeed waits to receive his enemy, but with the determination to fall upon him offensively at the proper moment, to wrest from him and transfer to his own troops the moral effect always produced by an onward movement when coupled with the certainty of throwing the main strength into the action at the most important point,—a thing altogether impossible when keeping strictly on the defensive. In fact, a general who occupies a well-chosen position, where his movements are free, has the advantage of observing the enemy's approach; his forces, previously arranged in a suitable manner upon the position, aided by batteries placed so as to produce the greatest effect, may make the enemy pay very dearly for his advance over the space separating the two armies; and when the assailant, after suffering severely, finds himself strongly assailed at the moment when the victory seemed to be in his hands, the advantage will, in all probability, be his no longer, for the moral effect of such a counter-attack upon the part of an adversary supposed to be beaten is certainly enough to stagger the boldest troops.

A general may, therefore, employ in his battles with equal success either the offensive or defensive system; but it is indispensable,—1st, that, so far from limiting himself to a passive defense, he should know how to take the offensive at favorable moments; 2d, that his *coup-d'oeil* be certain and his coolness undoubted; 3d, that he be able to rely surely upon his troops; 4th, that, in retaking the offensive, he should by no means neglect to apply the general principle which would have regulated his order of battle had he done so in the beginning; 5th, that he strike his blows upon decisive points. These truths are demonstrated by Napoleon's course at Rivoli and Austerlitz, as well as by Wellington's at Talavera, at Salamanca, and at Waterloo.

ARTICLE XXXI

OFFENSIVE BATTLES, AND DIFFERENT ORDERS OF BATTLE

We understand by offensive battles those which an army fights when assaulting another in position.[1] An army reduced to the strategic defensive often takes the offensive by making an attack, and an army receiving an attack may, during the progress of the battle, take the offensive and obtain the advantages incident to it. History furnishes numerous examples of battles of each of these kinds. As defensive battles have been discussed in the preceding article, and the advantages of the defensive been pointed out, we will now proceed to the consideration of offensive movements.

---

[1]   In every battle one party must be the assailant and the other assailed. Every battle is hence offensive for one party and defensive for the other.

It must be admitted that the assailant generally has a moral advantage over the assailed, and almost always acts more understandingly than the latter, who must be more or less in a state of uncertainty.

As soon as it is determined to attack the enemy, some order of attack must be adopted; and that is what I have thought ought to be called *order of battle*.

It happens also quite frequently that a battle must be commenced without a detailed plan, because the position of the enemy is not entirely known. In either case it should be well understood that there is in every battle-field a decisive point, the possession of which, more than of any other, helps to secure the victory, by enabling its holder to make a proper application of the principles of war: arrangements should therefore be made for striking the decisive blow upon this point.

The decisive point of a battle-field is determined, as has been already stated, by the character of the position, the bearing of different localities upon the strategic object in view, and, finally, by the arrangement of the contending forces. For example, suppose an enemy's flank to rest upon high ground from which his whole line might be attained, the occupation of this height seems most important, tactically considered; but it may happen that the height in question is very difficult of access, and situated exactly so as to be of the least importance, strategically considered. At the battle of Bautzen the left of the allies rested upon the steep mountains of Bohemia, which province was at that time rather neutral than hostile: it seemed that, tactically considered, the slope of these mountains was the decisive point to be held, when it was just the reverse, because the allies had but one line of retreat upon Reichenbach and Gorlitz, and the French, by forcing the right, which was in the plain, would occupy this line of retreat and throw the allies into the mountains, where they might have lost all their *materiel* and a great part of the personnel of their army. This course was also easier for them on account of the difference in the features of the ground, led to more important results, and would have diminished the obstacles in the future.

The following truths may, I think, be deduced from what has been stated: 1. The topographical key of a battle-field is not always the tactical key; 2. The decisive point of a battle-field is certainly that which combines strategic with topographical advantages; 3. When the difficulties of the ground are not too formidable upon the strategic point of the battle-field, this is generally the most important point; 4. It is nevertheless true that the determination of this point depends very much upon the arrangement of the contending forces. Thus, in lines of battle too much extended and divided the center will always be the proper point of attack; in lines well closed and connected the center is the strongest point, since, independently of the reserves posted there, it is easy to support it from the flanks: the decisive point in this case is therefore one of the extremities of the line. When the numerical superiority is considerable, an attack may be made simultaneously upon both extremities, but not when the attacking force is equal or inferior numerically to the enemy's. It appears, therefore, that all the combinations of a battle consist in so employing the force in hand as to obtain the most effective action upon that one of the three points mentioned which offers the greatest number of chances of success,—a point very easily determined by applying the analysis just mentioned.

The object of an offensive battle can only be to dislodge the enemy or to cut his line, unless it is intended by strategic maneuvers to ruin his army completely. An enemy is dislodged either by overthrowing him at some point of his line, or by outflanking him so as to take him in flank and rear, or by using both these methods at once; that is, attacking him in front while at the same time one wing is enveloped and his line turned.

To accomplish these different objects, it becomes necessary to make choice of the most suitable order of battle for the method to be used.

At least twelve orders of battle may be enumerated, viz.: 1. The simple parallel order; 2. The parallel order with a defensive or offensive crotchet; 3. The order reinforced upon one or both wings; 4. The order reinforced in the center; 5. The simple oblique order, or the oblique reinforced on the attacking wing; 6 and 7. The perpendicular order on one or both wings; 8. The concave order; 9. The convex order; 10. The order by echelon on one or both wings; 11. The order by echelon on the center; 12. The order resulting from a strong combined attack upon the center and one extremity simultaneously. (See Figs. 5 to 16.)

Figure 5[1]

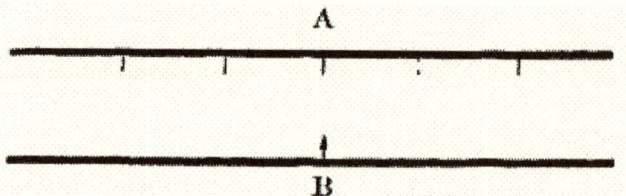

Each of these orders may be used either by itself or, as has been stated, in connection with the maneuver of a strong column intended to turn the enemy's line. In order to a proper appreciation of the merits of each, it becomes necessary to test each by the application of the general principles which have been laid down. For example, it is manifest that the parallel order (Fig. 5) is worst of all, for it requires no skill to fight one line against another, battalion against battalion, with equal chances of success on either side: no tactical skill is needed in such a battle.

There is, however, one important case where this is a suitable order, which occurs when an army, having taken the initiative in great strategic operations, shall have succeeded in falling upon the enemy's communications and cutting off his line of retreat while covering its own; when the battle takes place

[1] The letter A in this and other figures of the twelve orders indicates the defensive army, and B the offensive. The armies are represented each in a single line, in order not to complicate the figures too much; but it should be observed that every order of battle ought to be in two lines, whether the troops are deployed in columns of attack, in squares, or checkerwise.

between them, that army which has reached the rear of the other may use the parallel order, for, having effected the decisive maneuver previous to the battle, all its efforts should now be directed toward the frustration of the enemy's endeavor to open a way through for himself. Except for this single case, the parallel order is the worst of all. I do not mean to say that a battle cannot be gained while using this order, for one side or the other must gain the victory if the contest is continued; and the advantage will then be upon his side who has the best troops, who best knows when to engage them, who best manages his reserve and is most favored by fortune.

Figure 6.

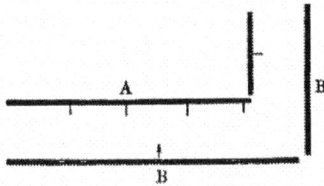

The parallel order with a crotchet upon the flank (Fig. 6) is most usually adopted in a defensive position. It may be also the result of an offensive combination; but then the crotchet is to the front, whilst in the case of defense it is to the rear. The battle of Prague is a very remarkable example of the danger to which such a crotchet is exposed if properly attacked.

Figure 7

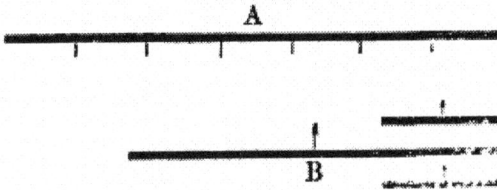

The parallel order reinforced upon one wing, (Fig. 7,) or upon the center, (Fig. 8) to pierce that of the enemy, is much more favorable than the two preceding ones, and is also much more in accordance with the general principles which have been laid down; although, when the contending forces are about equal, the part of the line which has been weakened to reinforce the other may have its own safety compromised if placed in line parallel to the enemy.

Figure 8

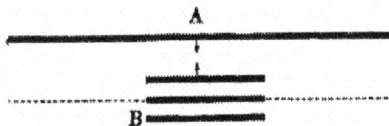

Figure 9

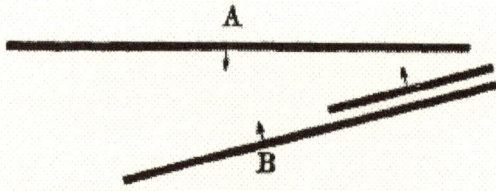

The oblique order (Fig. 9) is the best for an inferior force attacking a superior; for, in addition to the advantage of bringing the main strength of the forces against a single point of the enemy's line, it has two others equally important, since the weakened wing is not only kept back from the attack of the enemy, but performs also the double duty of holding in position the part of his line not attacked, and of being at hand as a reserve for the support, if necessary, of the engaged wing. This order was used by the celebrated Epaminondas at the battles of Leuctra and Mantinea. The most brilliant example of its use in modern times was given by Frederick the Great at the battle of Leuthen. (See Chapter VII. of Treatise on Grand Operations.)

Figure 10

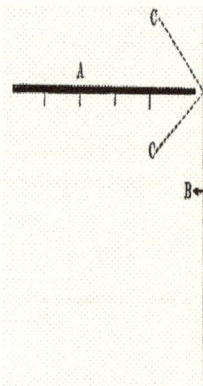

Figure 11

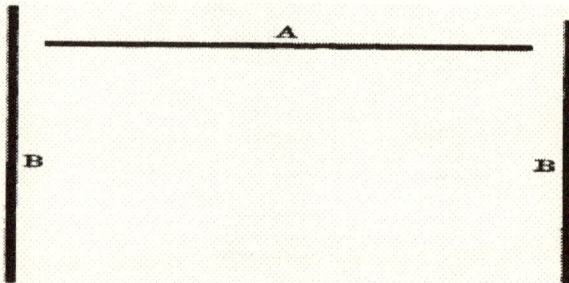

The perpendicular order on one or both wings, as seen in Figs. 10 and 11, can only be considered an arrangement to indicate the direction along which the primary tactical movements might be made in a battle. Two armies will never long occupy the relative perpendicular positions indicated in these figures; for if the army B were to take its first position on a line perpendicular to one or both extremities of the army A, the latter would at once change the front of a portion of its line; and even the army B, as soon as it extended itself to or beyond the extremity of A, must of necessity turn its columns either to the right or the left, in order to bring them near the enemy's line, and so take him in reverse, as at C, the result being two oblique lines, as shown in Fig. 10. The inference is that one division of the assailing army would take a position perpendicular to the enemy's wing, whilst the remainder of the army would approach in front for the purpose of annoying him; and this would always bring us back to one of the oblique orders shown in Figures 9 and 16.

The attack on both wings, whatever be the form of attack adopted, may be very advantageous, but it is only admissible when the assailant is very decidedly superior in numbers; for, if the fundamental principle is to bring the main strength of the forces upon the decisive point, a weaker army would violate it in directing a divided attack against a superior force. This truth will be clearly demonstrated farther on.

Figure 12

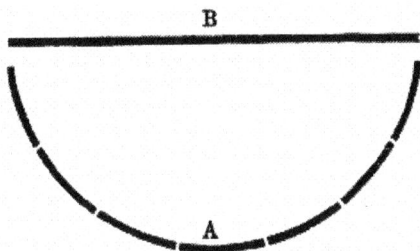

The order concave in the center (Fig. 12) has found advocates since the day when Hannibal by its use gained the battle of Cannae. This order may indeed be very good when the progress of the battle itself gives rise to it; that is, when the enemy attacks the center, this retires before him, and he suffers himself to be enveloped by the wings. But, if this order is adopted before the battle begins, the enemy, instead of falling on the center, has only to attack the wings, which present their extremities and are in precisely the same relative situation as if they had been assailed in flank. This order would, therefore, be scarcely ever used except against an enemy who had taken the convex order to fight a battle, as will be seen farther on.

Figure 12, *bis*

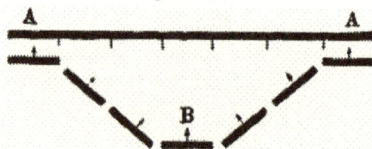

An army will rarely form a semicircle, preferring rather a broken line with the center retired, (Fig. 12, *bis*.) If several writers may be believed, such an arrangement gave the victory to the English on the famous days of Crecy and Agincourt. This order is certainly better than a semicircle, since it does not so much present the flank to attack, whilst allowing forward movement by echelon and preserving all the advantages of concentration of fire. These advantages vanish if the enemy, instead of foolishly throwing himself upon the retired center, is content to watch it from a distance and makes his greatest effort upon one wing. Essling, in 1809, is an example of the advantageous use of a concave line; but it must not be inferred that Napoleon committed an error in attacking the center; for an army fighting with the Danube behind it and with no way of moving without uncovering its bridges of communication, must not be judged as if it had been free to maneuver at pleasure.

Figure 13

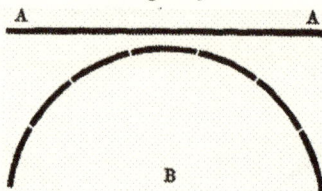

The convex order with the center salient (Fig. 13) answers for an engagement immediately upon the passage of a river when the wings must be retired and rested on the river to cover the bridges; also when a defensive battle is to be fought with a river in rear, which is to be passed and the defile covered, as at Leipsic; and, finally, it may become a natural formation to resist an enemy forming a concave line. If an enemy directs his efforts against the center or against a single wing, this order might cause the ruin of the whole army.[1]

The French tried it at Fleurus in 1794, and were successful, because the Prince of Coburg, in place of making a strong attack upon the center or upon a single extremity, divided his attack upon five or six diverging lines, and particularly upon both wings at once. Nearly the same convex order was adopted at Essling,

---

I   An attack upon the two extremities might succeed also in some cases, either when the force was strong enough to try it, or the enemy was unable to weaken his center to support the wings. As a rule, a false attack to engage the center, and a strong attack against one extremity, would be the best method to use against such a line.

and during the second and third days of the famous battle of Leipsic. On the last occasion it had just the result that might have been expected.

Figure 14

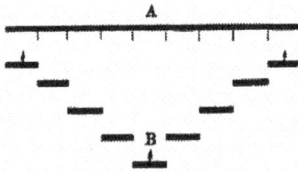

The order by echelon upon the two wings Fig. 14 is of the same nature as the perpendicular order, (Fig. 11,) being, however, better than that, because, the echelons being nearest each other in the direction where the reserve would be placed, the enemy would be less able, both as regards room and time, to throw himself into the interval of the center and make at that point a threatening counter-attack.

Figure 15

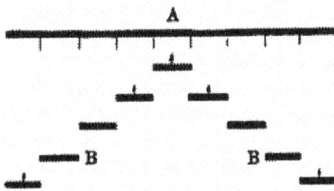

The order by echelon on the center (Fig. 15) may be used with special success against an army occupying a position too much cut up and too extended, because, its center being then somewhat isolated from the wings and liable to overthrow, the army thus cut in two would be probably destroyed. But, applying the test of the same fundamental principle, this order of attack would appear to be less certain of success against an army having a connected and closed line; for the reserve being generally near the center, and the wings being able to act either by concentrating their fire or by moving against the foremost echelons, might readily repulse them.

If this formation to some extent resembles the famous triangular wedge or *boar's head* of the ancients, and the column of Winkelried, it also differs from them essentially; for, instead of forming one solid mass,—an impracticable thing in our day, on account of the use of artillery,—it would have a large open space in the middle, which would render movements more easy. This formation is suitable, as has been said, for penetrating the center of a line too much extended, and might be equally successful against a line unavoidably immovable; but if the wings of the attacked line are brought at a proper time against the flanks of the foremost echelons, disagreeable consequences might result. A parallel order considerably reinforced on the center might perhaps be a much better arrangement, (Figs. 8 and 16;) for the parallel line in this case would have at least the advantage of deceiving the enemy as to the point of attack, and would hinder the wings from taking the echelons of the center by the flank.

This order by echelons was adopted by Laudon for the attack of the intrenched camp of Buntzelwitz. (Treatise on Grand Operations, chapter xxviii.) In such a case it is quite suitable; for it is then certain that the defensive army being forced to remain within its intrenchments, there is no danger of its attacking the echelons in flank. But, this formation having the inconvenience of indicating to the enemy the point of his line which it is desired to attack, false attacks should be made upon the wings, to mislead him as to the true point of attack.

Figure 16

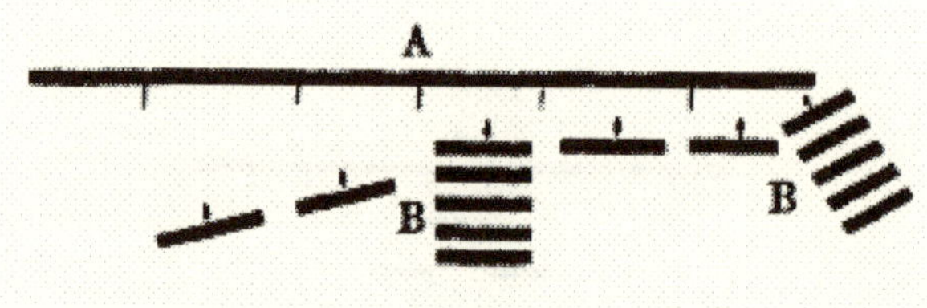

The order of attack in columns on the center and on one extremity at the same time (Fig. 16) is better than the preceding, especially in an attack upon an enemy's line strongly arranged and well connected. It may even be called the most reasonable of all the orders of battle. The attack upon the center, aided by a wing outflanking the enemy, prevents the assailed party falling upon the assailant and taking him in flank, as was done by Hannibal and Marshal Saxe. The enemy's wing which is hemmed in between the attacks on the center and at the extremity, having to contend with nearly the entire opposing force, will be defeated and probably destroyed. It was this maneuver which gave Napoleon his victories of Wagram and Ligny. This was what he wished to attempt at Borodino,—where he obtained only a partial success, on account of the heroic conduct of the Russian left and the division of Paskevitch in the famous central redoubt, and on account of the arrival of Baggavout's corps on the wing he hoped to outflank. He used it also at Bautzen,—where an unprecedented success would have been the result, but for an accident which interfered with the maneuver of the left wing intended to cut off the allies from the road to Wurschen, every arrangement having been made with that view.

It should be observed that these different orders are not to be understood precisely as the geometrical figures indicate them. A general who would expect to arrange his line of battle as regularly as upon paper or on a drill-ground would be greatly mistaken, and would be likely to suffer defeat. This is particularly true as battles are now fought. In the time of Louis XIV. or of Frederick, it was possible to form lines of battle almost as regular as the geometrical figures, because armies camped under tents, almost always closely collected together, and were in presence of each other several days, thus giving ample time for opening roads and clearing spaces to enable the columns to be at regular distances from each other. But in our day,—when armies bivouac, when their division into several corps gives greater mobility, when they take position near each other in obedience to orders given them while out of reach of the general's eye, and often when there has been no time for thorough examination of the enemy's position,—finally, when the different arms of the service are intermingled in the line of battle,—under these circumstances, all orders of battle which must be laid out with great accuracy of detail are impracticable. These figures have never been of any other use than to indicate approximate arrangements.

If every army were a solid mass, capable of motion as a unit under the influence of one man's will and as rapidly as thought, the art of winning battles would be reduced to choosing the most favorable order of battle, and a general could reckon with certainty upon the success of maneuvers arranged beforehand. But the facts are altogether different; for the great difficulty of the tactics of battles will always be to render certain the simultaneous entering into action of the numerous fractions whose efforts must combine to make such an attack as will give good ground to hope for victory: in other words, the chief difficulty is to cause these fractions to unite in the execution of the decisive maneuver which, in accordance with the original plan of the battle, is to result in victory.

Inaccurate transmission of orders, the manner in which they will be understood and executed by the subordinates of the general-in-chief, excess of activity in some, lack of it in others, a defective *coup-d'oeil militaire*,—every thing of this kind may interfere with the simultaneous entering into action of the different parts, without speaking of the accidental circumstances which may delay or prevent the arrival of a corps at the appointed place.

Hence result two undoubted truths: 1. The more simple a decisive maneuver is, the more sure of success will it be; 2. Sudden maneuvers seasonably executed during an engagement are more likely to succeed than those determined upon in advance, unless the latter, relating to previous strategic movements, will bring up the columns which are to decide the day upon those points where their presence will secure the expected result. Waterloo and Bautzen are proofs of the last. From the moment when Bluecher and Bulow had reached the heights of Frichermont, nothing could have prevented the loss of the battle by the French, and they could then only fight to make the defeat less complete. In like manner, at Bautzen, as soon as Ney had reached Klix, the retreat of the allies during the night of the 20th of May could alone have saved them, for on the 21st it was too late; and, if Ney had executed better what he was advised to do, the victory would have been a very great one.

As to maneuvers for breaking through a line and calculations upon the co-operation of columns proceeding from the general front of the army, with the intention of effecting large detours around an enemy's flank, it may be stated that their result is always doubtful, since it depends upon such an accurate execution of carefully-arranged plans as is rarely seen. This subject will be considered in Art. XXXII.

Besides the difficulty of depending upon the exact application of an order of battle arranged in advance, it often happens that battles begin without even the assailant having a well-defined object, although the collision may have been expected. This uncertainty results either from circumstances prior to the battle, from ignorance of the enemy's position and plans, or from the fact that a portion of the army may be still expected to arrive on the field.

From these things many people have concluded that it is impossible to reduce to different systems the formations of orders of battle, or that the adoption of either of them can at all influence the result of an engagement,—an erroneous conclusion, in my opinion, even in the cases cited above. Indeed, in battles begun without any predetermined plan it is probable that at the opening of the engagement the armies will occupy lines nearly parallel and more or less strengthened upon some point; the party acting upon the defensive, not knowing in what quarter the storm will burst upon him, will hold a large part of his forces in reserve, to be used as occasion may require; the assailant must make similar efforts to have his forces well in hand; but as soon as the point of attack shall have been determined, the mass of his troops will be directed against the center or upon one wing of the enemy, or upon both at once. Whatever may be the resulting formation, it will always bear a resemblance to one of the figures previously exhibited. Even in unexpected engagements the same thing would happen,—which will, it is hoped, be a sufficient proof of the fact that this classification of the different systems or orders of battle is neither fanciful nor useless.

There is nothing even in Napoleon's battles which disproves my assertion, although they are less susceptible than any others of being represented by lines accurately laid down. We see him, however, at Rivoli, at Austerlitz, and at Ratisbon, concentrating his forces toward the center to be ready at the favorable moment to fall upon the enemy. At the Pyramids he formed an oblique line of squares in echelon. At Leipsic, Essling, and Brienne he used a kind of convex order very like Fig. 11. At Wagram his order was altogether like Fig. 16, bringing up two masses upon the center and right, while keeping back the left wing; and this he wished to repeat at Borodino and at Waterloo before the Prussians came up. At Eylau, although the collision was almost entirely unforeseen on account of the very unexpected return and offensive movement of the Russians, he outflanked their left almost perpendicularly, whilst in another direction he was endeavoring to break through the center; but these attacks were not simultaneous, that on the center being repulsed at eleven o'clock, whilst Davoust did not attack vigorously upon the left until toward one. At Dresden he attacked by the two wings, for the first time probably in his life, because his center was covered by a fortification and an intrenched camp, and, in addition, the attack of his left was combined with that of Vandamme upon the enemy's line of retreat. At Marengo, if we may credit Napoleon himself, the oblique order he assumed, resting his right at Castel Ceriole, saved him from almost inevitable defeat. Ulm and Jena were battles won by strategy before they were fought, tactics having but little to do with them. At Ulm there was not even a regular battle.

I think we may hence conclude that if it seems absurd to desire to mark out upon the ground orders of battle in such regular lines as would be used in tracing them on a sketch, a skillful general may nevertheless bear in mind the orders which have been indicated above, and may so combine his troops on the battle-field that the arrangement shall be similar to one of them. He should endeavor in all his combinations, whether deliberately arranged or adopted on the spur of the moment, to form a sound conclusion as to the important point of the battle-field; and this he can only do by observing well the direction of the enemy's line of battle, and not forgetting the direction in which strategy requires him to operate. He will then give his attention and efforts to this point, using a third of his force to keep the enemy in check or watch his movements, while throwing the other two-thirds upon the point the possession of which will insure him the victory. Acting thus, he will have satisfied all the conditions the science of grand tactics can impose upon him, and will have applied the principles of the art in the most perfect manner. The manner of determining the decisive point of a battle-field has been described in the preceding chapter, (Art. XIX.)

Having now explained the twelve orders of battle, it has occurred to me that this would be a proper place to reply to several statements made in the Memoirs of Napoleon published by General Montholon.

The great captain seems to consider the oblique order a modern invention, a theorist's fancy,—an opinion I can by no means share; for the oblique order is as old as Thebes and Sparta, and I have seen it used with my own eyes. This assertion of Napoleon's seems the more remarkable because Napoleon himself boasted of having used, at Marengo, the very order of which he thus denies the existence.

If we understand that the oblique order is to be applied in the rigid and precise manner inculcated by General Ruchel at the Berlin school. Napoleon was certainly right in regarding it as an absurdity; but I repeat that a line of battle never was a regular geometrical figure, and when such figures are used in discussing the combinations of tactics it can only be for the purpose of giving definite expression to an idea by the use of a known symbol. It is nevertheless true that every line of battle which is neither parallel nor perpendicular to the enemy's must be oblique of necessity. If one army attacks the extremity of another army, the attacking wing being reinforced by massing troops upon it while the weakened wing is kept retired from attack, the direction of the line must of necessity be a little oblique, since one end of it will be nearer the enemy than the other. The oblique order is so far from being a mere fancy that we see it used when the order is that by echelons on one wing, (Fig. 14.)

As to the other orders of battle explained above, it cannot be denied that at Essling and Fleurus the general arrangement of the Austrians was a concave line, and that of the French a convex. In these orders parallel lines may be used as in the case of straight lines, and they would be classified as belonging to the parallel system when no part of the line was more strongly occupied or drawn up nearer to the enemy than another.

Laying aside for the present further consideration of these geometrical figures, it is to be observed that, for the purpose of fighting battles in a truly scientific manner, the following points must be attended to:–

1. An offensive order of battle should have for its object to force the enemy from his position by all reasonable means.

2. The maneuvers indicated by art are those intended to overwhelm one wing only, or the center and one wing at the same time. An enemy may also be dislodged by maneuvers for outflanking and turning his position.

3. These attempts have a much greater probability of success if concealed from the enemy until the very moment of the assault.

4. To attack the center and both wings at the same time, without having very superior forces, would be entirely in opposition to the rules of the art, unless one of these attacks can be made very strongly without weakening the line too much at the other points.

5. The oblique order has no other object than to unite at least half the force of the army in an overwhelming attack upon one wing, while the remainder is retired to the rear, out of danger of attack, being arranged either in echelon or in a single oblique line.

6 The different formations, convex, concave, perpendicular, or otherwise, may all be varied by having the lines of uniform strength throughout, or by massing troops at one point.

7. The object of the defense being to defeat the plans of the attacking party, the arrangements of a defensive order should be such as to multiply the difficulties of approaching the position, and to keep in hand a strong reserve, well concealed, and ready to fall at the decisive moment upon a point where the enemy least expect to meet it.

8. It is difficult to state with precision what is the best method to use in forcing a hostile army to abandon its position. An order of battle would be perfect which united the double advantages of the fire of the arms and of the moral effect produced by an onset. A skillful mixture of deployed lines and columns, acting alternately as circumstances require, will always be a good combination. In the practical use of this system many variations must arise from differences in the *coup-d'oeil* of commanders, the *morale* of officers and

soldiers, their familiarity with maneuvers and firings of all sorts, from varying localities, &c.

9. As it is essential in an offensive battle to drive the enemy from his position and to cut him up as much as possible, the best means of accomplishing this is to use as much material force as can be accumulated against him. It sometimes happens, however, that the direct application of main force is of doubtful utility, and better results may follow from maneuvers to outflank and turn that wing which is nearest the enemy's line of retreat. He may when thus threatened retire, when he would fight strongly and successfully if attacked by main force.

History is full of examples of the success of such maneuvers, especially when used against generals of weak character; and, although victories thus obtained are generally less decisive and the hostile army is but little demoralized, such incomplete successes are of sufficient importance not to be neglected, and a skillful general should know how to employ the means to gain them when opportunity offers, and especially should he combine these turning movements with attacks by main force.

10. The combination of these two methods—that is to say, the attack in front by main force and the turning maneuver—will render the victory more certain than the use of either separately; but, in all cases, too extended movements must be avoided, even in presence of a contemptible enemy.

11. The manner of driving an enemy from his position by main force is the following:–Throw his troops into confusion by a heavy and well-directed fire of artillery, increase this confusion by vigorous charges of cavalry, and follow up the advantages thus gained by pushing forward masses of infantry well covered in front by skirmishers and flanked by cavalry.

But, while we may expect success to follow such an attack upon the first line, the second is still to be overcome, and, after that, the reserve; and at this period of the engagement the attacking party would usually be seriously embarrassed, did not the moral effect of the defeat of the first line often occasion the retreat of the second and cause the general in command to lose his presence of mind. In fact, the attacking troops will usually be somewhat disordered, even in victory, and it will often be very difficult to replace them by those of the second line, because they generally follow the first line at such a distance as not to come within musket-range of the enemy; and it is always embarrassing to substitute one division for another in the heat of battle, at the moment when the enemy is putting forth all his strength in repelling the attack.

These considerations lead to the belief that if the general and the troops of the defensive army are equally active in the performance of their duty, and preserve their presence of mind, if their flanks and line of retreat are not threatened, the advantage will usually be on their side at the second collision of the battle; but to insure that result their second line and the cavalry must be launched against the victorious battalions of the adversary at the proper instant; for the loss of a few minutes may be irreparable, and the second line may be drawn into the confusion of the first.

12. From the preceding facts may be deduced the following truth: "that the most difficult as well as the most certain of all the means the assailant may use to gain the victory consists in strongly supporting the first line with the troops of the second line, and these with the reserve, and in a proper employment of masses of cavalry and of batteries, to assist in striking the decisive blow at the second line of the enemy; for here is presented the greatest of all the problems of the tactics of battles."

In this important crisis of battles, theory becomes an uncertain guide; for it is then unequal to the emergency, and can never compare in value with a natural talent for war, nor be a sufficient substitute for that intuitive *coup-d'oeil* imparted by experience in battles to a general of tried bravery and coolness.

The simultaneous employment of the largest number of troops of all arms combined, except a small reserve of each which should be always held in hand,[1] will, therefore, at the critical moment of the battle, be the problem which every skillful general will attempt to solve and to which he should give his whole attention. This critical moment is usually when the first line of the parties is broken, and all the efforts of both contestants are put forth,—on the one side to complete the victory, on the other to wrest it from the enemy. It is scarcely necessary to say that, to make this decisive blow more certain and effectual, a simultaneous attack upon the enemy's flank would be very advantageous.

13. In the defensive the fire of musketry can be much more effectively used than in the offensive, since when a position is to be carried it can be accomplished only by moving upon it, and marching and firing at the same time can be done only by troops as skirmishers, being an impossibility for the principal masses. The object of the defense being to break and throw into confusion the troops advancing to the attack, the fire of artillery and musketry will be the natural defensive means of the first line, and when the enemy presses too closely the columns of the second line and part of the cavalry must be launched against him. There will then be a strong probability of his repulse.

ARTICLE XXXII

TURNING MANEUVERS, AND TOO EXTENDED MOVEMENT IN BATTLES

We have spoken in the preceding article of maneuvers undertaken to turn an enemy's line upon the battle-field, and of the advantages which may be expected from them. A few words remain to be said as to the wide detours

---

[1] The great reserves must, of course, be also engaged when it is necessary; but it is always a good plan to keep back, as a final reserve, two or three battalions and five or six squadrons. Moreau decided the battle of Engen with four companies of infantry; and what Kellermann's cavalry accomplished at Marengo is known to every reader of history.

which these maneuvers sometimes occasion, causing the failure of so many plans seemingly well arranged.

It may be laid down as a principle that any movement is dangerous which is so extended as to give the enemy an opportunity, while it is taking place, of beating the remainder of the army in position. Nevertheless, as the danger depends very much upon the rapid and certain *coup-d'oeil* of the opposing general, as well as upon the style of warfare to which he is accustomed, it is not difficult to understand why so many maneuvers of this kind have failed against some commanders and succeeded against others, and why such a movement which would have been hazardous in presence of Frederick, Napoleon, or Wellington might have entire success against a general of limited capacity, who had not the tact to take the offensive himself at the proper moment, or who might himself have been in the habit of moving in this manner.

It seems, therefore, difficult to lay down a fixed rule on the subject. The following directions are all that can be given. Keep the mass of the force well in hand and ready to act at the proper moment, being careful, however, to avoid the danger of accumulating troops in too large bodies. A commander observing these precautions will be always prepared for any thing that may happen. If the opposing general shows little skill and seems inclined to indulge in extended movements, his adversary may be more daring.

A few examples drawn from history will serve to convince the reader of the truth of my statements, and to show him how the results of these extended movements depend upon the characters of the generals and the armies concerned in them.

In the Seven Years' War, Frederick gained the battle of Prague because the Austrians had left a feebly-defended interval of one thousand yards between their right and the remainder of their army,—the latter part remaining motionless while the right was overwhelmed. This inaction was the more extraordinary as the left of the Austrians had a much shorter distance to pass over in order to support their right than Frederick had to attack it; for the right was in the form of a crotchet, and Frederick was obliged to move on the arc of a large semicircle to reach it.

On the other hand, Frederick came near losing the battle of Torgau, because he made with his left a movement entirely too extended and disconnected (nearly six miles) with a view of turning the right of Marshal Daun.[1] Mollendorf brought up the right by a concentric movement to the heights of Siptitz, where he rejoined the king, whose line was thus reformed.

The battle of Rivoli is a noted instance in point. All who are familiar with that battle know that Alvinzi and his chief of staff Weyrother wished to surround Napoleon's little army, which was concentrated on the plateau of Rivoli. Their center was beaten,—while their left was piled up in the ravine of the Adige,

---

[1]   For an account of these two battles, see Chapters II. and XXV. of the Treatise on Grand Military Operations.

and Lusignan with their right was making a wide *detour* to get upon the rear of the French army, where he was speedily surrounded and captured.

No one can forget the day of Stockach, where Jourdan conceived the unfortunate idea of causing an attack to be made upon a united army of sixty thousand men by three small divisions of seven thousand or eight thousand men, separated by distances of several leagues, whilst Saint-Cyr, with the third of the army, (thirteen thousand men,) was to pass twelve miles beyond the right flank and get in rear of this army of sixty thousand men, which could not help being victorious over these divided fractions, and should certainly have captured the part in their rear. Saint-Cyr's escape was indeed little less than a miracle.

We may call to mind how this same General Weyrother, who had desired to surround Napoleon at Rivoli, attempted the same maneuver at Austerlitz, in spite of the severe lesson he had formerly received. The left wing of the allied army, wishing to outflank Napoleon's right, to cut him off from Vienna, (where he did not desire to return,) by a circular movement of nearly six miles, opened an interval of a mile and a half in their line. Napoleon took advantage of this mistake, fell upon the center, and surrounded their left, which was completely shut up between Lakes Tellnitz and Melnitz.

Wellington gained the battle of Salamanca by a maneuver very similar to Napoleon's, because Marmont, who wished to cut off his retreat to Portugal, left an opening of a mile and a half in his line,—seeing which, the English general entirely defeated his left wing, that had no support.

If Weyrother had been opposed to Jourdan at Rivoli or at Austerlitz, he might have destroyed the French army, instead of suffering in each case a total defeat; for the general who at Stockach attacked a mass of sixty thousand men with four small bodies of troops so much separated as to be unable to give mutual aid would not have known how to take proper advantage of a wide detour effected in his presence. In the same way, Marmont was unfortunate in having at Salamanca an adversary whose chief merit was a rapid and practiced tactical *coup-d'oeil*. With the Duke of York or Moore for an antagonist, Marmont would probably have been successful.

Among the turning maneuvers which have succeeded in our day, Waterloo and Hohenlinden had the most brilliant results. Of these the first was almost altogether a strategic operation, and was attended with a rare concurrence of fortunate circumstances. As to Hohenlinden, we will search in vain in military history for another example of a single brigade venturing into a forest in the midst of fifty thousand enemies, and there performing such astonishing feats as Richepanse effected in the defile of Matenpoet, where he might have expected, in all probability, to lay down his arms.

At Wagram the turning wing under Davoust contributed greatly to the successful issue of the day; but, if the vigorous attack upon the center under Macdonald, Oudinot, and Bernadotte had not rendered opportune assistance, it is by no means certain that a like success would have been the result.

So many examples of conflicting results might induce the conclusion that no rule on this subject can be given; but this would be erroneous; for it seems, on the contrary, quite evident that, by adopting as a rule an order of battle well closed and well connected, a general will find himself prepared for any emergency, and little will be left to chance; but it is specially important for him to have a correct estimate of his enemy's character and his usual style of warfare, to enable him to regulate his own actions accordingly. In case of superiority in numbers or discipline, maneuvers may be attempted which would be imprudent were the forces equal or the commanders of the same capacity. A maneuver to outflank and turn a wing should be connected with other attacks, and opportunely supported by an attempt of the remainder of the army on the enemy's front, either against the wing turned or against the center. Finally, strategic operations to cut an enemy's line of communications before giving battle, and attack him in rear, the assailing army preserving its own line of retreat, are much more likely to be successful and effectual, and, moreover, they require no disconnected maneuver during the battle.

ARTICLE XXXIII

UNEXPECTED MEETING OF TWO ARMIES ON THE MARCH

The accidental and unexpected meeting of two armies on the march gives rise to one of the most imposing scenes in war.

In the greater number of battles, one party awaits his enemy in a position chosen in advance, which is attacked after a reconnoissance as close and accurate as possible. It often happens, however,—especially as war is now carried on,—that two armies approach each other, each intending to make an unexpected attack upon the other. A collision ensues unexpected by both armies, since each finds the other where it does not anticipate a meeting. One army may also be attacked by another which has prepared a surprise for it,—as happened to the French at Rossbach.

A great occasion of this kind calls into play all the genius of a skillful general and of the warrior able to control events. It is always possible to gain a battle with brave troops, even where the commander may not have great capacity; but victories like those of Lutzen, Luzzara, Eylau, Abensberg, can only be gained by a brilliant genius endowed with great coolness and using the wisest combinations.

There is so much chance in these accidental battles that it is by no means easy to lay down precise rules concerning them; but these are the very cases in which it is necessary to keep clearly before the mind the fundamental principles of the art and the different methods of applying them, in order to a proper arrangement of maneuvers that must be decided upon at the instant and in the midst of the crash of resounding arms.

Two armies marching, as they formerly did, with all their camp-equipage, and meeting unexpectedly, could do nothing better at first than cause their advanced guard to deploy to the right or left of the roads they are traversing. In each army the forces should at the same time be concentrated so that they may be thrown in a proper direction considering the object of the march. A grave error would be committed in deploying the whole army behind the advanced guard; because, even if the deployment were accomplished, the result would be nothing more than a badly-arranged parallel order, and if the enemy pressed the advanced guard with considerable vigor the consequence might be the rout of the troops which were forming. (See the account of the battle of Rossbach, Treatise on Grand Operations.)

In the modern system, when armies are more easily moved, marching upon several roads, and divided into masses which may act independently, these routs are not so much to be feared; but the principles are unchanged. The advanced guard must always be halted and formed, and then the mass of the troops concentrated in that direction which is best suited for carrying out the object of the march. Whatever maneuvers the enemy may then attempt, every thing will be in readiness to meet him.

ARTICLE XXXIV

OF SURPRISES OF ARMIES

I shall not speak here of surprises of small detachments,—the chief features in the wars of partisan or light troops, for which the light Russian and Turkish cavalry are so well adapted. I shall confine myself to an examination of the surprise of whole armies.

Before the invention of fire-arms, surprises were more easily effected than at present; for the reports of artillery and musketry firing are heard to so great a distance that the surprise of an army is now next to an impossibility, unless the first duties of field-service are forgotten and the enemy is in the midst of the army before his presence is known because there are no outposts to give the alarm. The Seven Years' War presents a memorable example in the surprise of Hochkirch. It shows that a surprise does not consist simply in falling upon troops that are sleeping or keeping a poor look-out, but that it may result from the combination of a sudden attack upon, and a surrounding of, one extremity of the army. In fact, to surprise an army it is not necessary to take it so entirely unawares that the troops will not even have emerged from their tents, but it is sufficient to attack it in force at the point intended, before preparations can be made to meet the attack.

As armies at the present day seldom camp in tents when on a march, prearranged surprises are rare and difficult, because in order to plan one it becomes necessary to have an accurate knowledge of the enemy's camp. At

Marengo, at Lutzen, and at Eylau there was something like a surprise; but this term should only be applied to an entirely unexpected attack. The only great surprise to be cited is the case of Taroutin, in 1812, where Murat was attacked and beaten by Benningsen. To excuse his imprudence, Murat pretended that a secret armistice was in force; but there was really nothing of the kind, and he was surprised through his own negligence.

It is evident that the most favorable manner of attacking an army is to fall upon its camp just before daybreak, at the moment when nothing of the sort is expected. Confusion in the camp will certainly take place; and, if the assailant has an accurate knowledge of the locality and can give a suitable tactical and strategic direction to the mass of his forces, he may expect a complete success, unless unforeseen events occur. This is an operation by no means to be despised in war, although it is rare, and less brilliant than a great strategic combination which renders the victory certain even before the battle is fought.

For the same reason that advantage should be taken of all opportunities for surprising an adversary, the necessary precautions should be used to prevent such attacks. The regulations for the government of any well-organized army should point out the means for doing the last.

ARTICLE XXXV

OF THE ATTACK BY MAIN FORCE OF FORTIFIED PLACES, INTRENCHED CAMPS OR LINES.—OF COUPS DE MAIN IN GENERAL

There are many fortified places which, although not regular fortresses, are regarded as secure against *coups de main*, but may nevertheless be carried by escalade or assault, or through breaches not altogether practicable, but so steep as to require the use of ladders or some other means of getting to the parapet.

The attack of a place of this kind presents nearly the same combinations as that of an intrenched camp; for both belong to the class of *coups de main*.

This kind of attack will vary with circumstances: 1st, with the strength of the works; 2d, with the character of the ground on which they are built; 3d, with the fact of their being isolated or connected; 4th, with the morale of the respective parties. History gives us examples of all of these varieties.

For examples, take the intrenched camps of Kehl, Dresden, and Warsaw, the lines of Turin and Mayence, the intrenchments of Feldkirch, Scharnitz, and Assiette. Here I have mentioned several cases, each with varying circumstances and results. At Kehl (1796) the intrenchments were better connected and better constructed than at Warsaw. There was, in fact, a *tete de pont* nearly equal to a permanent fortification; for the archduke thought himself obliged to besiege it in form, and it would have been extremely hazardous for him to make an open attack upon it. At Warsaw the works were isolated, but of considerable relief, and they had as a keep a large city surrounded by loopholed walls, armed and defended by a number of desperate men.

Dresden, in 1813, had for a keep a bastioned enceinte, one front of which, however, was dismantled and had no other parapet than such as was suited to a field-work. The camp proper was protected by simple redoubts, at considerable distances apart, very poorly built, the keep giving it its sole strength.[1]

At Mayence and at Turin there were continuous lines of circumvallation; but if in the first case they were strong, they were certainly not so at Turin, where upon one of the important points there was an insignificant parapet with a command of three feet, and a ditch proportionally deep. In the latter case, also, the lines were between two fires, as they were attacked in rear by a strong garrison at the moment when Prince Eugene assailed them from without. At Mayence the lines were attacked in front, only a small detachment having succeeded in passing around the right flank.

The tactical measures to be taken in the attack of field-works are few in number. If it seems probable that a work may be surprised if attacked a little before day, it is altogether proper to make the attempt; but if this operation may be recommended in case of an isolated work, it is by no means to be expected that a large army occupying an intrenched camp will permit itself to be surprised,—especially as the regulations of all services require armies to stand to their arms at dawn. As an attack by main force seems likely to be the method followed in this case, the following simple and reasonable directions are laid down:–

1. Silence the guns of the work by a powerful artillery-fire, which at the same time has the effect of discouraging the defenders.

2. Provide for the troops all the materials necessary (such as fascines and short ladders) to enable them to pass the ditch and mount the parapet.

3. Direct three small columns upon the work to be taken, skirmishers preceding them, and reserves being at hand for their support.

4. Take advantage of every irregularity of the ground to get cover for the troops, and keep them sheltered as long as possible.

5. Give detailed instructions to the principal columns as to their duties when a work shall have been carried, and as to the manner of attacking the troops occupying the camp. Designate the bodies of cavalry which are to assist in attacking those troops if the ground permits. When all these arrangements are made, there is nothing more to be done but to bring up the troops to the attack as actively as possible, while a detachment makes an attempt at the gorge. Hesitancy and delay in such a case are worse than the most daring rashness.

Those gymnastic exercises are very useful which prepare soldiers for escalades and passing obstacles; and the engineers may with great advantage give their

---

[1]   The number of defenders at Dresden the first day (August 25) was twenty-four thousand, the next day, sixty-five thousand, and the third day, more than one hundred thousand.

attention to providing means for facilitating the passage of the ditches of field-works and climbing their parapets.

Among all the arrangements in cases of this kind of which I have read, none are better than those for the assault of Warsaw and the intrenched camp of Mayence. Thielke gives a description of Laudon's dispositions for attacking the camp of Buntzelwitz, which, although not executed, is an excellent example for instruction. The attack of Warsaw may be cited as one of the finest operations of this sort, and does honor to Marshal Paskevitch and the troops who executed it. As an example not to be followed, no better can be given than the arrangements made for attacking Dresden in 1813.

Among attacks of this class may be mentioned the memorable assaults or escalades of Port Mahon in 1756, and of Berg-op-zoom in 1747,—both preceded by sieges, but still brilliant *coups de main*, since in neither case was the breach sufficiently large for a regular assault.

Continuous intrenched lines, although seeming to have a better interconnection than lines of detached works, are more easily carried, because they may be several leagues in extent, and it is almost impossible to prevent an enemy from breaking through them at some point. The capture of the lines of Mayence and Wissembourg, which are described in the History of the Wars of the Revolution, (Chapters XXI. and XXII.,) and that of the lines of Turin by Eugene of Savoy in 1706, are excellent lessons for study.

This famous event at Turin, which has been so often referred to, is so familiar to all readers that it is unnecessary to recall the details of it; but I cannot pass it by without remarking how easily the victory was bought and how little it should have been expected. The strategic plan was certainly admirable; and the march from the Adige through Piacenza to Asti by the right bank of the Po, leaving the French on the Mincio, was beautifully arranged, but its execution was exceedingly slow. When we examine the operations near Turin, we must confess that the victors owed more to their good fortune than to their wisdom. It required no great effort of genius upon the part of Prince Eugene to prepare the order he issued to his army; and he must have felt a profound contempt for his opponents to execute a march with thirty-five thousand allied troops of ten different nations between eighty thousand Frenchmen on the one side and the Alps on the other, and to pass around their camp for forty-eight hours by the most remarkable flank march that was ever attempted. The order for the attack was so brief and so devoid of instruction that any staff officer of the present day ought to write a better. Directing the formation of eight columns of infantry by brigade in two lines, giving them orders to carry the intrenchments and to make openings through them for the passage of the cavalry into the camp, make up the sum total of all the science exhibited by Eugene in order to carry out his rash undertaking It is true he selected the weak point of the intrenchment; for it was there so low that it covered only half the bodies of its defenders.

But I am wandering from my subject, and must return to the explanation of the measures most suitable for adoption in an attack on lines. If they have a sufficient relief to make it difficult to carry them by assault, and if on the other hand they may be outflanked or turned by strategic maneuvers, it is far better to pursue the course last indicated than to attempt a hazardous assault. If, however, there is any reason for preferring the attack by assault, it should be made upon one of the wings, because the center is the point most easily succored. There have been cases where an attack on the wing was expected by the defenders, and they have been deceived by a false attack made at that point, while the real attack took place at the center, and succeeded simply because unexpected. In these operations the locality and the character of the generals engaged must decide as to the proper course to be pursued.

The attack may be executed in the manner described for intrenched camps. It has sometimes happened, however, that these lines have had the relief and proportions of permanent works; and in this case escalade would be quite difficult, except of old earthen works whose slopes were worn away from the lapse of time and had become accessible for infantry of moderate activity. The ramparts of Ismail and Praga were of this character; so also was the citadel of Smolensk, which Paskevitch so gloriously defended against Ney, because he preferred making his stand at the ravines in front, rather than take shelter behind a parapet with an inclination of scarcely thirty degrees.

If one extremity of a line rests upon a river, it seems absurd to think of penetrating upon that wing, because the enemy collecting his forces, the mass of which would be near the center, might defeat the columns advancing between the center and the river and completely destroy them. This absurdity, however, has sometimes been successful; because the enemy driven behind his lines rarely thinks of making an offensive return upon the assailant, no matter how advantageous it might seem. A general and soldiers who seek refuge behind lines are already half conquered, and the idea of taking the offensive does not occur to them when their intrenchments are attacked. Notwithstanding these facts, I cannot advise such a course; and the general who would run such a risk and meet the fate of Tallard at Blenheim could have no just cause of complaint.

Very few directions can be given for the defense of intrenched camps and lines. The first is to be sure of having strong reserves placed between the center and each wing, or, to speak more accurately, on the right of the left wing and on the left of the right wing. With this arrangement succor can be easily and rapidly carried to a threatened point, which could not be done were there but one central reserve. It has been suggested that three reserves would not be too many if the intrenchment is very extensive; but I decidedly incline to the opinion that two are quite enough. Another recommendation may be given, and it is of great importance,—that the troops be made to understand they must by no means despair of finally defending a line which may be forced at one point; because, if a good reserve is at hand, it may take the offensive, attack the assailant, and succeed in driving him out of the work he may have supposed in his power.

COUPS DE MAIN

These are bold enterprises undertaken by a detachment of an army for the capture of posts of different strength or importance.[1] They partake of the nature both of surprises and attacks by main force, for both these methods may be employed in carrying an attempt of this sort to a successful issue. Although *coups de main* seem to be entirely tactical operations, their importance certainly depends on the relations of the captured posts to the strategic combinations in hand. It will become necessary, therefore, to say a few words with reference to coups de main in Article XXXVI., when speaking of detachments. However tiresome these repetitions may seem, I am obliged to state here the manner of executing such operations, as it is evidently a part of the subject of the attack of intrenchments.

I do not pretend to say that the rules of tactics apply to these operations; for their name, *coups de main*, implies that ordinary rules are not applicable to them. I desire only to call attention to them, and refer my readers to the different works, either historical or didactic, where they are mentioned.

I have previously stated that important results may often follow from these enterprises. The capture of Sizeboli in 1828, the unsuccessful attack of General Petrasch upon Kehl in 1796, the remarkable surprises of Cremona in 1702, of Gibraltar in 1704, and of Berg-op-zoom in 1814, as well as the escalades of Port Mahon and Badajos, give an idea of the different kinds of *coup de main*. Some are effected by surprise, others by open force. Skill, stratagems, boldness, on the part of the assailant, and fear excited among the assailed, are some of the things which have an influence upon the successful issue of *coups de main*.

As war is now waged, the capture of a post, however strong, is no longer of the same importance as formerly unless it has a direct influence upon the results of a great strategic operation.

The capture or destruction of a bridge defended by intrenchments, that of a large convoy, of a small fort closing important passes, like the two attacks which were made in 1799 upon the fort of Lucisteig in the Grisons; the capture of Leutasch and Scharnitz by Ney in 1805; finally, the capture of a post not even fortified, but used as a great depot of provisions and munitions much needed by the enemy;—such are the enterprises which will justify the risks to which a detachment engaging in them may be exposed.

Posts have been captured by filling up the ditches sometimes with fascines, sometimes with bags of wool; and manure has been used for the same purpose.

---

1 The distinction between the importance and the strength of a post must be observed; for it may be very strong and of very little importance, and vice aversa.

Ladders are generally necessary, and should always be prepared. Hooks have been used in the hands and attached to the shoes of soldiers, to help them in climbing rocky heights which commanded the intrenchment. An entrance was effected through the sewers at Cremona by Prince Eugene.

In reading such facts, we must draw from them not rules, but hints; for what has been done once may be done again.

COMMENTARY ON CHAPTER IV

The chapter on grand tactics, which Jomini describes as the tactical combinations made by the commander in chief of an army, is a gold mine for historians. Jomini is quite comprehensive in his description of the various formations an experienced 19th century commander would have in his bag of tricks; a separate article deals with the problems associated with tactical turning maneuvers. The primary emphasis in this chapter is given over to the offensive side of the field, as obviously a defensive general is most likely reacting to what his opponent is doing. Jomini, however, is careful to emphasize that a defense should never remain entirely passive, but at some point should attempt to wrest the initiative away from his attacker. The weakest part of this chapter is the short section on surprise. Jomini seemed to think that, since surprise was an unreliable although desirable element in war, he could not wholeheartedly recommend trying to surprise an opponent. He was so determined to eliminate chaos and confusion from the battlefield that he neglected to see how a general could put surprise to use as an ally.

CHAPTER V

OF SEVERAL MIXED OPERATIONS, WHICH ARE IN CHARACTER PARTLY STRATEGICAL AND PARTLY TACTICAL

ARTICLE XXXVI
OF DIVERSIONS AND GREAT DETACHMENTS

The operations of the detachments an army may send out have so important a bearing on the success of a campaign, that the duty of determining their strength and the proper occasions for them is one of the greatest and most delicate responsibilities imposed upon a commander. If nothing is more useful in war than a strong detachment opportunely sent out and having a good *ensemble* of operations with the main body, it is equally certain that no expedient is more dangerous when inconsiderately adopted. Frederick the Great regarded it as one of the essential qualities of a general to know how to make his adversary send out many detachments, either with the view of destroying them in detail or of attacking the main body during their absence.

The division of armies into numerous detachments has sometimes been carried to so great an extent, and with such poor results, that many persons now believe it better to have none of them. It is undoubtedly much safer and more agreeable for an army to be kept in a single mass; but it is a thing at times impossible or incompatible with gaining a complete or even considerable success. The essential point in this matter is to send out as few detachments as possible.

There are several kinds of detachments.

1. There are large corps dispatched to a distance from the zone of operations of the main army, in order to make diversions of greater or less importance.

2. There are large detachments made in the zone of operations to cover important points of this zone, to carry on a siege, to guard a secondary base, or to protect the line of operations if threatened.

3. There are large detachments made upon the front of operations, in face of the enemy, to act in concert with the main body in some combined operation.

4. There are small detachments sent to a distance to try the effect of surprise upon isolated points, whose capture may have an important bearing upon the general operations of the campaign.

I understand by diversions those secondary operations carried out at a distance from the principal zone of operations, at the extremities of a theater of war, upon the success of which it is sometimes foolishly supposed the whole campaign depends. Such diversions are useful in but two cases, the first of which arises when the troops thus employed cannot conveniently act elsewhere on account of their distance from the real theater of operations, and the second is that where such a detachment would receive strong support from the population among which it was sent,—the latter case belonging rather to political than military combinations. A few illustrative examples may not be out of place here.

The unfortunate results for the allied powers of the Anglo-Russian expedition to Holland, and of that of the Archduke Charles toward the end of the last century, (which have been referred to in Article XIX.,) are well known.

In 1805, Napoleon was occupying Naples and Hanover. The allies intended an Anglo-Russian army to drive him out of Italy, while the combined forces of England, Russia, and Sweden should drive him from Hanover, nearly sixty thousand men being designed for these two widely-separated points. But, while their troops were collecting at the two extremities of Europe, Napoleon ordered the evacuation of Naples and Hanover, Saint-Cyr hastened to effect a junction with Massena in the Frioul, and Bernadotte, leaving Hanover, moved up to take part in the operations of Ulm and Austerlitz. After these astonishing successes, Napoleon had no difficulty in retaking Naples and Hanover. This is an example of the failure of diversions. I will give an instance where such an operation would have been proper.

In the civil wars of 1793, if the allies had sent twenty thousand men to La Vendee, they would have accomplished much more than by increasing the numbers of those who were fighting fruitlessly at Toulon, upon the Rhine, and in Belgium. Here is a case where a diversion would have been not only very useful, but decisive.

It has already been stated that, besides diversions to a distance and of small bodies, large corps are often detached in the zone of operations of the main army.

If the employment of these large corps thus detached for secondary objects is more dangerous than the diversions above referred to, it is no less true that they are often highly proper and, it may be, indispensable.

These great detachments are chiefly of two kinds. The first are permanent corps which must be sometimes thrown out in a direction opposite to the main line of operations, and are to remain throughout a campaign. The second are corps temporarily detached for the purpose of assisting in carrying out some special enterprise.

Among the first should be especially enumerated those fractions of an army that are detached either to form the strategic reserve, of which mention has been made, or to cover lines of operation and retreat when the configuration of the theater of the war exposes them to attack. For example, a Russian army that wishes to cross the Balkan is obliged to leave a portion of its forces to observe Shumla, Routchouk, and the valley of the Danube, whose direction is perpendicular to its line of operations. However successful it may be, a respectable force must always be left toward Giurgevo or Krajova, and even on the right bank of the river toward Routchouk.

This single example shows that it is sometimes necessary to have a double strategic front, and then the detachment of a considerable corps must be made to offer front to a part of the enemy's army in rear of the main army. Other localities and other circumstances might be mentioned where this measure would be equally essential to safety. One case is the double strategic front of the Tyrol and the Frioul for a French army passing the Adige. On whichever side it may wish to direct its main column, a detachment must be left on the other front sufficiently strong to hold in check the enemy threatening to cut the line of communications. The third example is the frontier of Spain, which enables the Spaniards to establish a double front,—one covering the road to Madrid, the other having Saragossa or Galicia as a base. To whichever side the invading army turns, a detachment must be left on the other proportioned in magnitude to the enemy's force in that direction.

All that can be said on this point is that it is advantageous to enlarge as much as possible the field of operations of such detachments, and to give them as much power of mobility as possible, in order to enable them by opportune movements to strike important blows. A most remarkable illustration of this truth was given by Napoleon in the campaign of 1797. Obliged as he was to leave a corps of fifteen thousand men in the valley of the Adige to observe the Tyrol while he was operating toward the Noric Alps, he preferred to draw this corps to his aid, at the risk of losing temporarily his line of retreat, rather than leave the parts of his army disconnected and exposed to defeat in detail. Persuaded that he could be victorious with his army united, he apprehended no particular danger from the presence of a few hostile detachments upon his communications.

Great movable and temporary detachments are made for the following reasons:–

1. To compel your enemy to retreat to cover his line of operations, or else to cover your own.

2. To intercept a corps and prevent its junction with the main body of the enemy, or to facilitate the approach of your own reinforcements.

3. To observe and hold in position a large portion of the opposing army, while a blow is struck at the remainder.

4. To carry off a considerable convoy of provisions or munitions, on receiving which depended the continuance of a siege or the success of any strategic enterprise, or to protect the march of a convoy of your own.

5. To make a demonstration to draw the enemy in a direction where you wish him to go, in order to facilitate the execution of an enterprise in another direction.

6. To mask, or even to invest, one or more fortified places for a certain time, with a view either to attack or to keep the garrison shut up within the ramparts.

7. To take possession of an important point upon the communications of an enemy already retreating.

However great may be the temptation to undertake such operations as those enumerated, it must be constantly borne in mind that they are always secondary in importance, and that the essential thing is to be successful at the decisive points. A multiplication of detachments must, therefore, be avoided. Armies have been destroyed for no other reason than that they were not kept together.

We will here refer to several of these enterprises, to show that their success depends sometimes upon good fortune and sometimes upon the skill of their designer, and that they often fail from faulty execution.

Peter the Great took the first step toward the destruction of Charles XII. by causing the seizure, by a strong detachment, of the famous convoy Lowenhaupt was bringing up. Villars entirely defeated at Denain the large detachment Prince Eugene sent out in 1709 under D'Albermale.

The destruction of the great convoy Laudon took from Frederick during the siege of Olmutz compelled the king to evacuate Moravia. The fate of the two detachments of Fouquet at Landshut in 1760, and of Fink at Maxen in 1759, demonstrates how difficult it is at times to avoid making detachments, and how dangerous they may be. To come nearer our own times, the disaster of Vandamme at Culm was a bloody lesson, teaching that a corps must not be thrust forward too boldly: however, we must admit that in this case the operation was well planned, and the fault was not so much in sending out the detachment as in not supporting it properly, as might easily have been done. That of Fink was destroyed at Maxen nearly on the same spot and for the same reason.

Diversions or demonstrations in the zone of operations of the army are decidedly advantageous when arranged for the purpose of engaging the enemy's attention in one direction, while the mass of the forces is collected upon another point where the important blow is to be struck. In such a case, care must be taken not only to avoid engaging the corps making the demonstration, but to recall it promptly toward the main body. We will mention two examples as illustrations of these facts.

In 1800, Moreau, wishing to deceive Kray as to the true direction of his march, carried his left wing toward Rastadt from Kehl, whilst he was really filing off his army toward Stockach; his left, having simply shown itself, returned toward the center by Fribourg in Brisgau.

In 1805, Napoleon, while master of Vienna, detached the corps of Bernadotte to Iglau to overawe Bohemia and paralyze the Archduke Ferdinand, who was assembling an army in that territory; in another direction he sent Davoust to Presburg to show himself in Hungary; but he withdrew them to Brunn, to take part in the event which was to decide the issue of the campaign, and a great and decisive victory was the result of his wise maneuvers. Operations of this kind, so far from being in opposition to the principles of the art of war, are necessary to facilitate their application.

It readily appears from what goes before that precise rules cannot be laid down for these operations, so varied in character, the success of which depends on so many minute details. Generals should run the risk of making detachments only after careful consideration and observation of all the surrounding circumstances. The only reasonable rules on the subject are these: send out as few detachments as possible, and recall thorn immediately when their duty is performed. The inconveniences necessarily attending them may be made as few as practicable, by giving judicious and carefully-prepared instructions to their commanders: herein lies the great talent of a good chief of staff.

One of the means of avoiding the disastrous results to which detachments sometimes lead is to neglect none of the precautions prescribed by tactics for increasing the strength of any force by posting it in good positions; but it is generally imprudent to engage in a serious conflict with too large a body of troops. In such cases ease and rapidity of motion will be most likely to insure safety. It seldom happens that it is right for a detachment to resolve to conquer or die in the position it has taken, whether voluntarily or by order.

It is certain that in all possible cases the rules of tactics and of field-fortification must be applied by detachments as well as by the army itself.

Since we have included in the number of useful cases of detachments those intended for *coups de main*, it is proper to mention a few examples of this kind to enable the reader to judge for himself. We may call to mind that one which was executed by the Russians toward the end of 1828 with the view of taking possession of Sizeboli in the Gulf of Bourghas. The capture of this feebly-fortified gulf, which the Russians rapidly strengthened, procured for them in case of success an essential *point d'appui* beyond the Balkan, where depots could be established in advance for the army intending to cross those mountains: in case of failure, no one was compromised,—not even the small corps which had been debarked, since it had a safe and certain retreat to the shipping.

Chapter V. Of Several Mixed Operations, Which Are in
Character Partly Strategical and Partly Tactical

In like manner, in the campaign of 1796, the *coup de main* attempted by
the Austrians for the purpose of taking possession of Kehl and destroying
the bridge whilst Moreau was returning from Bavaria, would have had very
important consequences if it had not failed.

In attempts of this kind a little is risked to gain a great deal; and, as they
can in no wise compromise the safety of the main army, they may be freely
recommended.

Small bodies of troops thrown forward into the zone of the enemy's
operations belong to the class of detachments that are judicious. A few hundred
horsemen thus risked will be no great loss if captured; and they may be the
means of causing the enemy great injury. The small detachments sent out by
the Russians in 1807, 1812, and 1813 were a great hinderance to Napoleon's
operations, and several times caused his plans to fail by intercepting his
couriers.

For such expeditions officers should be selected who are bold and full of
stratagems. They ought to inflict upon the enemy all the injury they can
without compromising themselves. When an opportunity of striking a telling
blow presents itself, they should not think for a moment of any dangers or
difficulties in their path. Generally, however, address and presence of mind,
which will lead them to avoid useless danger, are qualities more necessary
for a partisan than cool, calculating boldness. For further information on
this subject I refer my readers to Chapter XXXV. of the Treatise on Grand
Operations, and to Article XLV. of this work, on light cavalry.

ARTICLE XXXVII

PASSAGE OF RIVERS AND OTHER STREAMS

The passage of a small stream, over which a bridge is already in place or might
be easily constructed, presents none of the combinations belonging to grand
tactics or strategy; but the passage of a large river, such as the Danube, the
Rhine, the Po, the Elbe, the Oder, the Vistula, the Inn, the Ticino, &c, is an
operation worthy the closest study.

The art of building military bridges is a special branch of military science,
which is committed to pontoniers or sappers. It is not from this point of view
that I propose to consider the passage of a stream, but as the attack of a military
position and as a maneuver.

The passage itself is a tactical operation; but the determination of the point
of passage may have an important connection with all the operations taking
place within the entire theater of the war. The passage of the Rhine by General
Moreau in 1800 is an excellent illustration of the truth of this remark. Napoleon,
a more skillful strategist than Moreau, desired him to cross at Schaffhausen in
order to take Kray's whole army in reverse, to reach Ulm before him, to cut him

off from Austria and hurl him back upon the Main. Moreau, who had already a bridge at Basel, preferred passing, with greater convenience to his army, in front of the enemy, to turning his extreme left. The tactical advantages seemed to his mind much more sure than the strategical: he preferred the certainty of a partial success to the risk attending a victory which would have been a decisive one. In the same campaign Napoleon's passage of the Po is another example of the high strategic importance of the choice of the point of crossing. The army of the reserve, after the engagement of the Chiusella, could either march by the left bank of the Po to Turin, or cross the river at Crescentino and march directly to Genoa. Napoleon preferred to cross the Ticino, enter Milan, effect a junction with Moncey who was approaching with twenty thousand men by the Saint-Gothard pass, then to cross the Po at Piacenza, expecting to get before Melas more certainly in that direction than if he came down too soon upon his line of retreat. The passage of the Danube at Donauwerth and Ingolstadt in 1805 was a very similar operation. The direction chosen for the passage was the prime cause of the destruction of Mack's army.

The proper strategic point of passage is easily determined by recollecting the principles laid down in Article XIX.; and it is here only necessary to remind the reader that in crossing a river, as in every other operation, there are permanent or geographical decisive points, and others which are relative or eventual, depending on the distribution of the hostile forces.

If the point selected combines strategic advantages with the tactical, no other point can be better; but if the locality presents obstacles exceedingly difficult to pass, another must be chosen, and in making the new selection care should be taken to have the direction of the movement as nearly as possible coincident with the true strategic direction. Independently of the general combinations, which exercise a great influence in fixing the point of passage, there is still another consideration, connected with the locality itself. The best position is that where the army after crossing can take its front of operations and line of battle perpendicular to the river, at least for the first marches, without being forced to separate into several corps moving upon different lines. This advantage will also save it the danger of fighting a battle with a river in rear, as happened to Napoleon at Essling.

Enough has been said with reference to the strategical considerations influencing the selection of the point of crossing a river. We will now proceed to speak of the passage itself. History is the best school in which to study the measures likely to insure the success of such operations. The ancients deemed the passage of the Granicus—which is a small stream—a wonderful exploit. So far as this point is concerned, the people of modern days can cite much greater.

The passage of the Rhine at Tholhuys by Louis XIV. has been greatly lauded; and it was really remarkable. In our own time, General Dedon has made famous the two passages of the Rhine at Kehl and of the Danube at Hochstadt in 1800. His work is a model as far as concerns the details; and in these operations minute attention to details is every thing. More recently, three other passages of the Danube, and the ever-famous passage of the Beresina, have exceeded every thing of the kind previously seen. The two first were executed by Napoleon at Essling and at Wagram, in presence of an army of one hundred and twenty thousand men provided with four hundred pieces of cannon, and at a point where the bed of the stream is broadest. General Pelet's interesting account of them should be carefully read. The third was executed by the Russian army at Satounovo in 1828, which, although not to be compared with the two just mentioned, was very remarkable on account of the great local difficulties and the vigorous exertions made to surmount them. The passage of the Beresina was truly wonderful. My object not being to give historical details on this subject, I direct my readers to the special narratives of these events. I will give several general rules to be observed.

1. It is essential to deceive the enemy as to the point of passage, that he may not accumulate an opposing force there. In addition to the strategic demonstrations, false attacks must be made near the real one, to divide the attention and means of the enemy. For this purpose half of the artillery should be employed to make a great deal of noise at the points where the passage is not to be made, whilst perfect silence should be preserved where the real attempt is to be made.

2. The construction of the bridge should be covered as much as possible by troops sent over in boats for the purpose of dislodging the enemy who might interfere with the progress of the work; and these troops should take possession at once of any villages, woods, or other obstacles in the vicinity.

3. It is of importance also to arrange large batteries of heavy caliber, not only to sweep the opposite bank, but to silence any artillery the enemy might bring up to batter the bridge while building. For this purpose it is convenient to have the bank from which the passage is made somewhat higher than the other.

4. The proximity of a large island near the enemy's bank gives great facilities for passing over troops in boats and for constructing the bridge. In like manner, a smaller stream emptying into the larger near the point of passage is a favorable place for collecting and concealing boats and materials for the bridge.

5. It is well to choose a position where the river makes a re-entering bend, as the batteries on the assailant's side can cross their fire in front of the point where the troops are to land from the boats and where the end of the bridge is to rest, thus taking the enemy in front and flank when he attempts to oppose the passage.

6. The locality selected should be near good roads on both banks, that the army may have good communications to the front and rear on both banks of the river. For this reason, those points where the banks are high and steep should be usually avoided.

The rules for preventing a passage follow as a matter of course from those for effecting it, as the duty of the defenders is to counteract the efforts of the assailants. The important thing is to have the course of the river watched by bodies of light troops, without attempting to make a defense at every point. Concentrate rapidly at the threatened point, in order to overwhelm the enemy while a part only of his army shall have passed. Imitate the Duke of Vendome at Cassano, and the Archduke Charles at Essling in 1809,—the last example being particularly worthy of praise, although the operation was not so decidedly successful as might have been expected.

In Article XXI. attention was called to the influence that the passage of a river, in the opening of a campaign, may have in giving direction to the lines of operations. We will now see what connection it may have with subsequent strategic movements.

One of the greatest difficulties to be encountered after a passage is to cover the bridge against the enemy's efforts to destroy it, without interfering too much with the free movement of the army. When the army is numerically very superior to the enemy, or when the river is passed just after a great victory gained, the difficulty mentioned is trifling; but when the campaign is just opening, and the two opposing armies are about equal, the case is very different.

If one hundred thousand Frenchmen pass the Rhine at Strasbourg or at Manheim in presence of one hundred thousand Austrians, the first thing to be done will be to drive the enemy in three directions,—first, before them as far as the Black Forest, secondly, by the right in order to cover the bridges on the Upper Rhine, and thirdly, by the left to cover the bridges of Mayence and the Lower Rhine. This necessity is the cause of an unfortunate division of the forces; but, to make the inconveniences of this subdivision as few as possible, the idea must be insisted on that it is by no means essential for the army to be separated into three equal parts, nor need these detachments remain absent longer than the few days required for taking possession of the natural point of concentration of the enemy's forces.

The fact cannot be concealed, however, that the case supposed is one in which the general finds his position a most trying one; for if he divides his army to protect his bridges he may be obliged to contend with one of his subdivisions against the whole of the enemy's force, and have it overwhelmed; and if he moves his army upon a single line, the enemy may divide his army and reassemble it at some unexpected point, the bridges may be captured or destroyed, and the general may find himself compromised before he has had time or opportunity to gain a victory.

The best course to be pursued is to place the bridges near a city which will afford a strong defensive point for their protection, to infuse all possible vigor and activity into the first operations after the passage, to fall upon the subdivisions of the enemy's army in succession, and to beat them in such a way that they will have no further desire of touching the bridges. In some cases eccentric lines of operations may be used. If the enemy has divided his one hundred thousand men into several corps, occupying posts of observation, a passage may be effected with one hundred thousand men at a single point near the center of the line of posts, the isolated defensive corps at this position may be overwhelmed, and two masses of fifty thousand men each may then be formed, which, by taking diverging lines of operations, can certainly drive off the successive portions of the opposing army, prevent them from reuniting, and remove them farther and farther from the bridges. But if, on the contrary, the passage be effected at one extremity of the enemy's strategic front, by moving rapidly along this front the enemy may be beaten throughout its whole extent,—in the same manner that Frederick tactically beat the Austrian line at Leuthen throughout its length,—the bridges will be secure in rear of the army, and remain protected during all the forward movements. It was in this manner that Jourdan, having passed the Rhine at Dusseldorf in 1795, on the extreme right of the Austrians, could have advanced in perfect safety toward the Main. He was driven away because the French, having a double and exterior line of operations, left one hundred and twenty thousand men inactive between Mayence and Basel, while Clairfayt repulsed Jourdan upon the Lahn. But this cannot diminish the importance of the advantages gained by passing a river upon one extremity of the enemy's strategic front. A commander-in-chief should either adopt this method, or that previously explained, of a central mass at the moment of passage, and the use of eccentric lines afterward, according to the circumstances of the case, the situation of the frontiers and bases of operations, as well as the positions of the enemy. The mention of these combinations, of which something has already been said in the article on lines of operations, does not appear out of place here, since their connection with the location of bridges has been the chief point under discussion.

It sometimes happens that, for cogent reasons, a double passage is attempted upon a single front of operations, as was the case with Jourdan and Moreau in 1796. If the advantage is gained of having in case of need a double line of retreat, there is the inconvenience, in thus operating on the two extremities of the enemy's front, of forcing him, in a measure, to concentrate on his center, and he may be placed in a condition to overwhelm separately the two armies which have crossed at different points. Such an operation will always lead to disastrous results when the opposing general has sufficient ability to know how to take advantage of this violation of principles.

In such a case, the inconveniences of the double passage may be diminished by passing over the mass of the forces at one of the points, which then becomes the decisive one, and by concentrating the two portions by interior lines as rapidly as possible, to prevent the enemy from destroying them separately. If Jourdan and Moreau had observed this rule, and made a junction of their forces in the direction of Donauwerth, instead of moving eccentrically, they would probably have achieved great successes in Bavaria, instead of being driven back upon the Rhine.

ARTICLE XXXVIII

RETREATS AND PURSUITS

Retreats are certainly the most difficult operations in war. This remark is so true that the celebrated Prince de Ligne said, in his usual piquant style, that he could not conceive how an army ever succeeded in retreating. When we think of the physical and moral condition of an army in full retreat after a lost battle, of the difficulty of preserving order, and of the disasters to which disorder may lead, it is not hard to understand why the most experienced generals have hesitated to attempt such an operation.

What method of retreat shall be recommended? Shall the fight be continued at all hazards until nightfall and the retreat executed under cover of the darkness? or is it better not to wait for this last chance, but to abandon the field of battle while it can be done and a strong opposition still made to the pursuing army? Should a forced march be made in the night, in order to get as much start of the enemy as possible? or is it better to halt after a half-march and make a show of fighting again? Each of these methods, although entirely proper in certain cases, might in others prove ruinous to the whole army. If the theory of war leaves any points unprovided for, that of retreats is certainly one of them.

If you determine to fight vigorously until night, you may expose yourself to a complete defeat before that time arrives; and if a forced retreat must begin when the shades of night are shrouding every thing in darkness and obscurity, how can you prevent the disintegration of your army, which does not know what to do, and cannot see to do any thing properly? If, on the other hand, the field of battle is abandoned in broad daylight and before all possible efforts have been made to hold it, you may give up the contest at the very moment when the enemy is about to do the same thing; and this fact coming to the knowledge of the troops, you may lose their confidence,—as they are always inclined to blame a prudent general who retreats before the necessity for so doing may be evident to themselves. Moreover, who can say that a retreat commenced in the daylight in presence of an enterprising enemy may not become a rout?

When the retreat is actually begun, it is no less difficult to decide whether a forced march shall be made to get as much the start of the enemy as possible,—since this hurried movement might sometimes cause the destruction of the army, and might, in other circumstances, be its salvation. All that can be positively asserted on this subject is that, in general, with an army of considerable magnitude, it is best to retreat slowly, by short marches, with a well-arranged rear-guard of sufficient strength to hold the heads of the enemy's columns in check for several hours.

Retreats are of different kinds, depending upon the cause from which they result. A general may retire of his own accord before fighting, in order to draw his adversary to a position which he prefers to his present one. This is rather a prudent maneuver than a retreat. It was thus that Napoleon retired in 1805 from Wischau toward Brunn to draw the allies to a point which suited him as a battle-field. It was thus that Wellington retired from Quatre-Bras to Waterloo. This is what I proposed to do before the attack at Dresden, when the arrival of Napoleon was known. I represented the necessity of moving toward Dippoldiswalde to choose a favorable battle-field. It was supposed to be a retreat that I was proposing; and a mistaken idea of honor prevented a retrograde movement without fighting, which would have been the means of avoiding the catastrophe of the next day, (August 26, 1813.)

A general may retire in order to hasten to the defense of a point threatened by the enemy, either upon the flanks or upon the line of retreat. When an army is marching at a distance from its depots, in an exhausted country, it may be obliged to retire in order to get nearer its supplies. Finally, an army retires involuntarily after a lost battle, or after an unsuccessful enterprise.

These are not the only causes having an influence in retreats. Their character will vary with that of the country, with the distances to be passed over and the obstacles to be surmounted. They are specially dangerous in an enemy's country; and when the points at which the retreats begin are distant from the friendly country and the base of operations, they become painful and difficult.

From the time of the famous retreat of the Ten Thousand, so justly celebrated, until the terrible catastrophe which befell the French army in 1812, history does not make mention of many remarkable retreats. That of Antony, driven out of Media, was more painful than glorious. That of the Emperor Julian, harassed by the same Parthians, was a disaster. In more recent days, the retreat of Charles VIII. to Naples, when he passed by a corps of the Italian army at Fornovo, was an admirable one. The retreat of M. de Bellisle from Prague does not deserve the praises it has received. Those executed by the King of Prussia after raising the siege of Olmutz and after the surprise at Hochkirch were very well arranged; but they were for short distances. That of Moreau in 1796, which was magnified in importance by party spirit, was creditable, but not at all extraordinary. The retreat of Lecourbe from Engadin to Altorf, and that of Macdonald by Pontremoli after the defeat of the Trebbia, as also that of Suwaroff from the Muttenthal to Chur, were glorious feats of arms, but partial in character and of short duration. The retreat of the Russian army from the Niemen to Moscow—a space of two hundred and forty leagues,—in presence of such an enemy as Napoleon and such cavalry as the active and daring Murat commanded, was certainly admirable. It was undoubtedly attended by many favorable circumstances, but was highly deserving of praise, not only for the talent displayed by the generals who directed its first stages, but also for the admirable fortitude and soldierly bearing of the troops who performed it. Although the retreat from Moscow was a bloody catastrophe for Napoleon, it was also glorious for him and the troops who were at Krasnoi and the Beresina,—because the skeleton of the army was saved, when not a single man should have returned. In this ever-memorable event both parties covered themselves with glory.

The magnitude of the distances and the nature of the country to be traversed, the resources it offers, the obstacles to be encountered, the attacks to be apprehended, either in rear or in flank, superiority or inferiority in cavalry, the spirit of the troops, are circumstances which have a great effect in deciding the fate of retreats, leaving out of consideration the skillful arrangements which the generals may make for their execution.

A general falling back toward his native land along his line of magazines and supplies may keep his troops together and in good order, and may effect a retreat with more safety than one compelled to subsist his army in cantonments, finding it necessary to occupy an extended position. It would be absurd to pretend that a French army retiring from Moscow to the Niemen without supplies of provisions, in want of cavalry and draft horses, could effect the movement in the same good order and with the same steadiness as a Russian army, well provided with every thing necessary, marching in its own country, and covered by an immense number of light cavalry.

There are five methods of arranging a retreat:–

The first is to march in a single mass and upon one road.

The second consists in dividing the army into two or three corps, marching at the distance of a day's march from each other, in order to avoid confusion, especially in the *materiel*.

The third consists in marching upon a single front by several roads nearly parallel and having a common point of arrival.

The fourth consists in moving by constantly converging roads.

The fifth, on the contrary, consists in moving along diverging roads.

I have nothing to say as to the formation of rear-guards; but it is taken for granted that a good one should always be prepared and well sustained by a portion of the cavalry reserves. This arrangement is common to all kinds of retreats, but has nothing to do with the strategic relations of these operations.

An army falling back in good order, with the intention of fighting as soon as it shall have received expected reinforcements or as soon as it shall have reached a certain strategic position, should prefer the first method, as this particularly insures the compactness of the army and enables it to be in readiness for battle almost at any moment, since it is simply necessary to halt the heads of columns and form the remainder of the troops under their protection as they successively arrive. An army employing this method must not, however, confine itself to the single main road, if there are side-roads sufficiently near to be occupied which may render its movements more rapid and secure.

When Napoleon retired from Smolensk, he used the second method, having the portions of his army separated by an entire march. He made therein a great mistake, because the enemy was not following upon his rear, but moving along a lateral road which brought him in a nearly perpendicular direction into the midst of the separated French corps. The three fatal days of Krasnoi were the result. The employment of this method being chiefly to avoid incumbering the road, the interval between the departure of the several corps is sufficiently great when the artillery may readily file off. Instead of separating the corps by a whole march, the army would be better divided into two masses and a rear-guard, a half-march from each other. These masses, moving off in succession with an interval of two hours between the departure of their several army-corps, may file off without incumbering the road, at least in ordinary countries. In crossing the Saint-Bernard or the Balkan, other calculations would doubtless be necessary.

I apply this idea to an army of one hundred and twenty thousand or one hundred and fifty thousand men, having a rear-guard of twenty thousand or twenty-five thousand men distant about a half-march in rear. The army may be divided into two masses of about sixty thousand men each, encamped at a distance of three or four leagues from each other. Each of these masses will be subdivided into two or three corps, which may either move successively along the road or form in two lines across the road. In either case, if one corps of thirty thousand men moves at five A.M. and the other at seven, there will be no danger of interference with each other, unless something unusual should happen; for the second mass being at the same hours of the day about four leagues behind the first, they can never be occupying the same part of the road at the same time.

When there are practicable roads in the neighborhood, suitable at least for infantry and cavalry, the intervals may be diminished. It is scarcely necessary to add that such an order of march can only be used when provisions are plentiful; and the third method is usually the best, because the army is then marching in battle-order. In long days and in hot countries the best times for marching are the night and the early part of the day. It is one of the most difficult problems of logistics to make suitable arrangements of hours of departures and halts for armies; and this is particularly the case in retreats.

Many generals neglect to arrange the manner and times of halts, and great disorder on the march is the consequence, as each brigade or division takes the responsibility of halting whenever the soldiers are a little tired and find it agreeable to bivouac. The larger the army and the more compactly it marches, the more important does it become to arrange well the hours of departures and halts, especially if the army is to move at night. An ill-timed halt of part of a column may cause as much mischief as a rout.

If the rear-guard is closely pressed, the army should halt in order to relieve it by a fresh corps taken from the second mass, which will halt with this object in view. The enemy seeing eighty thousand men in battle-order will think it necessary to halt and collect his columns; and then the retreat should recommence at nightfall, to regain the space which has been lost.

The third method, of retreating along several parallel roads, is excellent when the roads are sufficiently near each other. But, if they are quite distant, one wing separated from the center and from the other wing may be compromised if the enemy attacks it in force and compels it to stand on the defensive. The Prussian army moving from Magdeburg toward the Oder, in 1806, gives an example of this kind.

The fourth method, which consists in following concentric roads, is undoubtedly the best if the troops are distant from each other when the retreat is ordered. Nothing can be better, in such a case, than to unite the forces; and the concentric retreat is the only method of effecting this.

The fifth method indicated is nothing else than the famous system of eccentric lines, which I have attributed to Bulow, and have opposed so warmly in the earlier editions of my works, because I thought I could not be mistaken either as to the sense of his remarks on the subject or as to the object of his system. I gathered from his definition that he recommended to a retreating army, moving from any given position, to separate into parts and pursue diverging roads, with the double object of withdrawing more readily from the enemy in pursuit and of arresting his march by threatening his flanks and his line of communications. I found great fault with the system, for the simple reason that a beaten army is already weak enough, without absurdly still further dividing its forces and strength in presence of a victorious enemy.

Bulow has found defenders who declare that I mistake his meaning, and that by the term *eccentric retreat* he did not understand a retreat made on several diverging roads, but one which, instead of being directed toward the center of the base of operations or the center of the country, should be eccentric to that focus of operations, and along the line of the frontier of the country.

I may possibly have taken an incorrect impression from his language, and in this case my criticism falls to the ground; for I have strongly recommended that kind of a retreat to which I have given the name of the parallel retreat. It is my opinion that an army, leaving the line which leads from the frontiers to the center of the state, with a view of moving to the right or the left, may very well pursue a course nearly parallel to the line of the frontiers, or to its front of operations and its base. It seems to me more rational to give the name of parallel retreat to such a movement as that described, designating as eccentric retreat that where diverging roads are followed, all leading from the strategic front.

However this dispute about words may result, the sole cause of which was the obscurity of Bulow's text, I find fault only with those retreats made along several diverging roads, under pretense of covering a greater extent of frontier and of threatening the enemy on both flanks.

By using these high-sounding words *flanks*, an air of importance may be given to systems entirely at variance with the principles of the art. An army in retreat is always in a bad state, either physically or morally; because a retreat can only be the result of reverses or of numerical inferiority. Shall such an army be still more weakened by dividing it? I find no fault with retreats executed in several columns, to increase the ease of moving, when these columns can support each other; but I am speaking of those made along diverging lines of operations. Suppose an army of forty thousand men retreating before another of sixty thousand. If the first forms four isolated divisions of about ten thousand men, the enemy may maneuver with two masses of thirty thousand men each. Can he not turn his adversary, surround, disperse, and ruin in succession all his divisions? How can they escape such a fate? *By concentration.* This being in direct opposition to a divergent system, the latter falls of itself.

I invoke to my support the great lessons of experience. When the leading divisions of the army of Italy were repulsed by Wurmser, Bonaparte collected them all together at Roverbella; and, although he had only forty thousand men, he fought and beat sixty thousand, because he had only to contend against isolated columns. If he had made a divergent retreat, what would have become of his army and his victories? Wurmser, after his first check, made an eccentric retreat, directing his two wings toward the extremities of the line of defense. What was the result? His right, although supported by the mountains of the Tyrol, was beaten at Trent. Bonaparte then fell upon the rear of his left, and destroyed that at Bassano and Mantua.

When the Archduke Charles gave way before the first efforts of the French armies in 1796, would he have saved Germany by an eccentric movement? Was not the salvation of Germany due to his concentric retreat? At last Moreau, who had moved with a very extended line of isolated divisions, perceived that this was an excellent system for his own destruction, if he stood his ground and fought or adopted the alternative of retreating. He concentrated his scattered troops, and all the efforts of the enemy were fruitless in presence of a mass which it was necessary to watch throughout the whole length of a line of two hundred miles. Such examples must put an end to further discussion.[1]

There are two cases in which divergent retreats are admissible, and then only as a last resource. First, when an army has experienced a great defeat in its own country, and the scattered fragments seek protection within the walls of fortified places. Secondly, in a war where the sympathies of the whole population are enlisted, each fraction of the army thus divided may serve as a nucleus of assembly in each province; but in a purely methodical war, with regular armies, carried on according to the principles of the art, divergent retreats are simply absurd.

There is still another strategical consideration as to the direction of a retreat,—to decide when it should be made perpendicularly to the frontier and toward the interior of the country, or when it should be parallel to the frontier. For example, when Marshal Soult gave up the line of the Pyrenees in 1814, he had to choose one of two directions for his retreat,—either by way of

---

1 Ten years after this first refutation of Bulow's idea, the concentric retreat of Barclay and Bagration saved the Russian army. Although it did not prevent Napoleon's first success, it was, in the end, the cause of his ruin.

Bordeaux toward the interior of France, or by way of Toulouse parallel to the frontier formed by the Pyrenees. In the same way, when Frederick retired from Moravia, he marched toward Bohemia instead of returning to Silesia.

These parallel retreats are often to be preferred, for the reason that they divert the enemy from a march upon the capital of the state and the center of its power. The propriety of giving such a direction to a retreat must be determined by the configuration of the frontiers, the positions of the fortresses, the greater or less space the army may have for its marches, and the facilities for recovering its direct communications with the central portions of the state.

Spain is admirably suited to the use of this system. If a French army penetrates by way of Bayonne, the Spaniards may base themselves upon Pampeluna and Saragossa, or upon Leon and the Asturias; and in either case the French cannot move directly to Madrid, because their line of operations would be at the mercy of their adversary.

The frontier of the Turkish empire on the Danube presents the same advantages, if the Turks knew how to profit by them.

In France also the parallel retreat may be used, especially when the nation itself is not divided into two political parties each of which is striving for the possession of the capital. If the hostile army penetrates through the Alps, the French can act on the Rhone and the Saone, passing around the frontier as far as the Moselle on one side, or as far as Provence on the other. If the enemy enters the country by way of Strasbourg, Mayence, or Valenciennes, the same thing can be done. The occupation of Paris by the enemy would be impossible, or at least very hazardous, so long as a French army remained in good condition and based upon its circle of fortified towns. The same is the case for all countries having double fronts of operations.[1]

Austria is perhaps not so fortunately situated, on account of the directions of the Rhetian and Tyrolean Alps and of the river Danube. Lloyd, however, considers Bohemia and the Tyrol as two bastions connected by the strong curtain of the river Inn, and regards this frontier as exceedingly well suited for parallel movements. This assertion was not well sustained by the events of the campaigns of 1800, 1805, and 1809; but, as the parallel method has not yet had a fair trial on that ground, the question is still an open one.

It seems to me that the propriety of applying the parallel method depends mainly upon the existing and the antecedent circumstances of each case. If a French army should approach from the Rhine by way of Bavaria, and should find allies in force upon the Lech and the Iser, it would be a very delicate operation to throw the whole Austrian army into the Tyrol and into Bohemia, with the expectation of arresting in this way the forward movement to Vienna. If half the Austrian army is left upon the Inn to cover the approaches to the capital, an unfortunate division of force is the consequence; and if it is decided

---

[1] In all these calculations I suppose the contending forces nearly equal. If the invading army is twice as strong as the defensive, it may be divided into two equal parts, one of which may move directly upon the capital, while the other may follow the army retiring along the frontier. If the armies are equal, this is impossible.

to throw the whole army into the Tyrol, leaving the way to Vienna open, there would be great danger incurred if the enemy is at all enterprising. In Italy, beyond the Mincio, the parallel method would be of difficult application on the side of the Tyrol, as well as in Bohemia against an enemy approaching from Saxony, for the reason that the theater of operations would be too contracted.

In Prussia the parallel retreat may be used with great advantage against an army debouching from Bohemia upon the Elbe or the Oder, whilst its employment would be impossible against a French army moving from the Rhine, or a Russian army from the Vistula, unless Prussia and Austria were allies. This is a result of the geographical configuration of the country, which allows and even favors lateral movements: in the direction of its greatest dimension, (from Memel to Mayence;) but such a movement would be disastrous if made from Dresden to Stettin.

When an army retreats, whatever may be the motive of the operation, a pursuit always follows.

A retreat, even when executed in the most skillful manner and by an army in good condition, always gives an advantage to the pursuing army; and this is particularly the case after a defeat and when the source of supplies and reinforcements is at a great distance; for a retreat then becomes more difficult than any other operation in war, and its difficulties increase in proportion to the skill exhibited by the enemy in conducting the pursuit.

The boldness and activity of the pursuit will depend, of course, upon the character of the commanders and upon the *physique* and *morale* of the two armies. It is difficult to prescribe fixed rules for all cases of pursuits, but the following points must be recollected:–

1. It is generally better to direct the pursuit upon the flank of the retreating columns, especially when it is made in one's own country and where no danger is incurred in moving perpendicularly or diagonally upon the enemy's line of operations. Care must, however, be taken not to make too large a circuit; for there might then be danger of losing the retreating enemy entirely.

2. A pursuit should generally be as boldly and actively executed as possible, especially when it is subsequent to a battle gained; because the demoralized army may be wholly dispersed if vigorously followed up.

3. There are very few cases where it is wise to make a bridge of gold for the enemy, no matter what the old Roman proverb may say; for it can scarcely ever be desirable to pay an enemy to leave a country, unless in the case when an unexpected success shall have been gained over him by an army much inferior to his in numbers.

Nothing further of importance can be added to what has been said on the subject of retreats, as far as they are connected with grand combinations of strategy. We may profitably indicate several tactical measures which may render them more easy of execution.

One of the surest means of making a retreat successfully is to familiarize the officers and soldiers with the idea that an enemy may be resisted quite as well when coming on the rear as on the front, and that the preservation of order is the only means of saving a body of troops harassed by the enemy during a retrograde movement. Rigid discipline is at all times the best preservative of good order, but it is of special importance during a retreat. To enforce discipline, subsistence must be furnished, that the troops may not be obliged to straggle off for the purpose of getting supplies by marauding.

It is a good plan to give the command of the rear-guard to an officer of great coolness, and to attach to it staff officers who may, in advance of its movements, examine and select points suitable for occupation to hold the enemy temporarily in check. Cavalry can rally so rapidly on the main body that it is evidently desirable to have considerable bodies of such troops, as they greatly facilitate the execution of a slow and methodical retreat, and furnish the means of thoroughly examining the road itself and the neighborhood, so as to prevent an unexpected onset of the enemy upon the flanks of the retreating columns.

It is generally sufficient if the rear-guard keep the enemy at the distance of half a day's march from the main body. The rear-guard would run great risk of being itself cut off, if farther distant. When, however, there are defiles in its rear which are held by friends, it may increase the sphere of its operations and remain a full day's march to the rear; for a defile, when held, facilitates a retreat in the same degree that it renders it more difficult if in the power of the enemy. If the army is very numerous and the rear-guard proportionally large, it may remain a day's march in rear. This will depend, however, upon its strength, the nature of the country, and the character and strength of the pursuing force. If the enemy presses up closely, it is of importance not to permit him to do so with impunity, especially if the retreat is made in good order. In such a case it is a good plan to halt from time to time and fall unexpectedly upon the enemy's advanced guard, as the Archduke Charles did in 1796 at Neresheim, Moreau at Biberach, and Kleber at Ukerath. Such a maneuver almost always succeeds, on account of the surprise occasioned by an unexpected offensive return upon a body of troops which is thinking of little else than collecting trophies and spoils.

Passages of rivers in retreat are also operations by no means devoid of interest. If the stream is narrow and there are permanent bridges over it, the operation is nothing more than the passage of a defile; but when the river is wide and is to be crossed upon a temporary military bridge, it is a maneuver of extreme delicacy. Among the precautions to be taken, a very important one is to get the parks well advanced, so that they may be out of the way of the army; for this purpose it is well for the army to halt a half-day's march from the river. The rear-guard should also keep at more than the usual distance from the main body,—as far, in fact, as the locality and the respective forces opposed will permit. The army may thus file across the bridge without being too much hurried. The march of the rear-guard should be so arranged that it shall have reached a position in front of the bridge just as the last of the main body has passed. This will be a suitable moment for relieving the rear-guard by fresh troops strongly posted. The rear-guard will pass through the intervals of the fresh troops in position and will cross the river; the enemy, coming up and finding fresh troops drawn up to give him battle, will make no attempt to press them too closely. The new rear-guard will hold its position until night, and will then cross the river, breaking the bridges after it.

It is, of course, understood that as fast as the troops pass they form on the opposite bank and plant batteries, so as to protect the corps left to hold the enemy in check.

The dangers of such a passage in retreat, and the nature of the precautions which facilitate it, indicate that measures should always be taken to throw up intrenchments at the point where the bridge is to be constructed and the passage made. Where time is not allowed for the construction of a regular *tete de pont*, a few well-armed redoubts will be found of great value in covering the retreat of the last troops.

If the passage of a large river is so difficult when the enemy is only pressing on the rear of the column, it is far more so when the army is threatened both in front and rear and the river is guarded by the enemy in force.

The celebrated passage of the Beresina by the French is one of the most remarkable examples of such an operation. Never was an army in a more desperate condition, and never was one extricated more gloriously and skillfully. Pressed by famine, benumbed with cold, distant twelve hundred miles from its base of operations, assailed by the enemy in front and in rear, having a river with marshy banks in front, surrounded by vast forests, how could it hope to escape? It paid dearly for the honor it gained. The mistake of Admiral Tschitchagoff doubtless helped its escape; but the army performed heroic deeds, for which due praise should be given. We do not know whether to admire most the plan of operations which brought up the Russian armies from the extremities of Moldavia, from Moscow, and from Polotzk to the Beresina as to a rendezvous arranged in peace,—a plan which came near effecting the capture of their formidable adversary,—or the wonderful firmness of the lion thus pursued, who succeeded in opening a way through his enemies.

The only rules to be laid down are, not to permit your army to be closely pressed upon, to deceive the enemy as to the point of passage, and to fall headlong upon the corps which bars the way before the one which is following the rear of your column can come up. Never place yourself in a position to be exposed to such danger; for escape in such a case is rare.

If a retreating army should strive to protect its bridges either by regular *tetes de pont*, or at least by lines of redoubts to cover the rear-guard, it is natural, also, that the enemy pursuing should use every effort to destroy the bridges. When the retreat is made down the bank of a river, wooden houses may be thrown into the stream, also fire-ships and mills,—a means the Austrians used in 1796 against Jourdan's army, near Neuwied on the Rhine, where they nearly compromised the army of the Sambre and the Meuse. The Archduke Charles did the same thing at Essling in 1809. He broke the bridge over the Danube, and brought Napoleon to the brink of ruin.

It is difficult to secure a bridge against attacks of this character unless there is time for placing a stockade above it. Boats may be anchored, provided with ropes and grappling-hooks to catch floating bodies and with means for extinguishing fire-boats.

ARTICLE XXXIX
OF CANTONMENTS, EITHER WHEN ON THE MARCH, OR WHEN ESTABLISHED IN WINTER QUARTERS

So much has been written on this point, and its connection with my subject is so indirect, that I shall treat it very briefly.

To maintain an army in cantonments, in a war actively carried on, is generally difficult, however connected the arrangement may be, and there is almost always some point exposed to the enemy's attacks. A country where large towns abound, as Lombardy, Saxony, the Netherlands, Swabia, or old Prussia, presents more facilities for the establishment of quarters than one where towns are few; for in the former case the troops have not only convenient supplies of food, but shelters which permit the divisions of the army to be kept closely together. In Poland, Russia, portions of Austria and France, in Spain and in Southern Italy, it is more difficult to put an army into winter quarters.

Formerly, it was usual for each party to go into winter quarters at the end of October, and all the fighting after that time was of a partisan character and carried on by the advanced troops forming the outposts.

The surprise of the Austrian winter quarters in Upper Alsace in 1674, by Turenne, is a good example, from which may be learned the best method of conducting such an enterprise, and the precautions to be taken on the other side to prevent its success.

The best rules to be laid down on this subject seem to me to be the following. Establish the cantonments very compactly and connectedly and occupying a space as broad as long, in order to avoid having a too extended line of troops, which is always easily broken through and cannot be concentrated in time; cover them by a river, or by an outer line of troops in huts and with their position strengthened by field-works; fix upon points of assembly which may be reached by all the troops before the enemy can penetrate so far; keep all the avenues by which an enemy may approach constantly patrolled by bodies of cavalry; finally, establish signals to give warning if an attack is made at any point.

In the winter of 1807, Napoleon established his army in cantonments behind the Passarge in face of the enemy, the advanced guard alone being hutted near the cities of Gutstadt, Osterode, &c. The army numbered more than one hundred and twenty thousand men, and much skill was requisite in feeding it and keeping it otherwise comfortable in this position until June. The country was of a favorable character; but this cannot be expected to be the case everywhere.

An army of one hundred thousand men may find it not very difficult to have a compact and well-connected system of winter quarters in countries where large towns are numerous. The difficulty increases with the size of the army. It must be observed, however, that if the extent of country occupied increases in proportion to the numbers in the army, the means of opposing an irruption of the enemy increase in the same proportion. The important point is to be able to assemble fifty thousand or sixty thousand men in twenty-four hours. With such an army in hand, and with the certainty of having it rapidly increased, the enemy may be held in check, no matter how strong he may be, until the whole army is assembled.

It must be admitted, however, that there will always be a risk in going into winter quarters if the enemy keeps his army in a body and seems inclined to make offensive movements; and the conclusion to be drawn from this fact is, that the only method of giving secure repose to an army in winter or in the midst of a campaign is to establish it in quarters protected by a river, or to arrange an armistice.

In the strategic positions taken up by an army in the course of a campaign, whether marching, or acting as an army of observation, or waiting for a favorable opportunity of taking the offensive, it will probably occupy quite compact cantonments. The selection of such positions requires great experience upon the part of a general, in order that he may form correct conclusions as to what he may expect the enemy to do. An army should occupy space enough to enable it to subsist readily, and it should also keep as much concentrated as possible, to be ready for the enemy should he show himself; and these two conditions are by no means easily reconciled. There is no better arrangement than to place the divisions of the army in a space nearly a square, so that in case of need the whole may be assembled at any point where the enemy may present himself. Nine divisions placed in this way, a half-day's march from each other, may in twelve hours assemble on the center. The same rules are to be observed in these cases as were laid down for winter quarters.

ARTICLE XL
DESCENTS

These are operations of rare occurrence, and may be classed as among the most difficult in war when effected in presence of a well-prepared enemy.

Since the invention of gunpowder and the changes effected by it in navies, transports are so helpless in presence of the monstrous three-deckers of the present day, armed as they are with a hundred cannon, that an army can make a descent only with the assistance of a numerous fleet of ships of war which can command the sea, at least until the debarkation of the army takes place.

Before the invention of gunpowder, the transports were also the ships of war; they were moved along at pleasure by using oars, were light, and could skirt along the coasts; their number was in proportion to the number of troops to be embarked; and, aside from the danger of tempests, the operations of a fleet could be arranged with almost as much certainty as those of an army on land. Ancient history, for these reasons, gives us examples of more extensive debarkations than modern times.

Who does not recall to mind the immense forces transported by the Persians upon the Black Sea, the Bosporus, and the Archipelago,—the innumerable hosts landed in Greece by Xerxes and Darius,—the great expeditions of the Carthaginians and Romans to Spain and Sicily, that of Alexander into Asia Minor, those of Caesar to England and Africa, that of Germanicus to the mouths of the Elbe,—the Crusades,—the expeditions of the Northmen to England, to France, and even to Italy?

Since the invention of cannon, the too celebrated Armada of Philip II. was the only enterprise of this kind of any magnitude until that set on foot by Napoleon against England in 1803. All other marine expeditions were of no great extent: as, for example, those of Charles V. and of Sebastian of Portugal to the coast of Africa; also the several descents of the French into the United States of America, into Egypt and St. Domingo, of the English to Egypt, Holland, Copenhagen, Antwerp, Philadelphia. I say nothing of Hoche's projected landing in Ireland; for that was a failure, and is, at the same time, an example of the difficulties to be apprehended in such attempts.

The large armies kept on foot in our day by the great states of the world prevent descents with thirty or forty thousand men, except against second-rate powers; for it is extremely difficult to find transportation for one hundred or one hundred and fifty thousand men with their immense trains of artillery, munitions, cavalry, &c.

We were, however, on the point of seeing the solution of the vast problem of the practicability of descents in great force, if it is true that Napoleon seriously contemplated the transportation of one hundred and sixty thousand veterans from Boulogne to the British Isles: unfortunately, his failure to execute this gigantic undertaking has left us entirely in the dark as to this grave question.

It is not impossible to collect fifty French ships-of-the-line in the Channel by misleading the English; this was, in fact, upon the point of being done; it is then no longer impossible, with a favorable wind, to pass over the flotilla in two days and effect a landing. But what would become of the army if a storm should disperse the fleet of ships of war and the English should return in force to the Channel and defeat the fleet or oblige it to regain its ports?

Posterity will regret, as the loss of an example to all future generations, that this immense undertaking was not carried through, or at least attempted. Doubtless, many brave men would have met their deaths; but were not those men mowed down more uselessly on the plains of Swabia, of Moravia, and of Castile, in the mountains of Portugal and the forests of Lithuania? What man would not glory in assisting to bring to a conclusion the greatest trial of skill and strength ever seen between two great nations? At any rate, posterity will find in the preparations made for this descent one of the most valuable lessons the present century has furnished for the study of soldiers and of statesmen. The labors of every kind performed on the coasts of France from 1803 to 1805 will be among the most remarkable monuments of the activity, foresight, and skill of Napoleon. It is recommended to the careful attention of young officers. But, while admitting the possibility of success for a great descent upon a coast so near as the English to Boulogne, what results should be expected if this armada had had a long sea-voyage to make? How could so many small vessels be kept moving, even for two days and nights? To what chances of ruin would not so many frail boats be exposed in navigating the open seas! Moreover, the artillery, munitions of war, equipments, provisions, and fresh water that must be carried with this multitude of men require immense labor in preparation and vast means of transportation.

Experience has shown clearly the difficulties attending such an expedition, even for thirty thousand men. From known facts, it is evident that a descent can be made with this number of men in four cases:—1st, against colonies or isolated possessions; 2d, against second-rate powers which cannot be immediately supported from abroad; 3d, for the purpose of effecting a temporary diversion, or to capture a position which it is important to hold for a time; 4th, to make a diversion, at once political and military, against a state already engaged in a great war, whose troops are occupied at a distance from the point of the descent.

It is difficult to lay down rules for operations of this character. About the only recommendations I can make are the following. Deceive the enemy as to the point of landing; choose a spot where the vessels may anchor in safety and the troops be landed together; infuse as much activity as possible into the operation, and take possession of some strong point to cover the development of the troops as they land; put on shore at once a part of the artillery, to give confidence and protection to the troops that have landed.

A great difficulty in such an operation is found in the fact that the transports can never get near the beach, and the troops must be landed in boats and rafts,—which takes time and gives the enemy great advantages. If the sea is rough, the men to be landed are exposed to great risks; for what can a body of infantry do, crowded in boats, tossed about by the waves, and ordinarily rendered unfit by sea-sickness for the proper use of their arms?

I can only advise the party on the defensive not to divide his forces too much by attempting to cover every point. It is an impossibility to line the entire coast with batteries and battalions for its defense; but the approaches to those places where large establishments are to be protected must be closed. Signals should be arranged for giving prompt notice of the point where the enemy is landing, and all the disposable force should be rapidly concentrated there, to prevent his gaining a firm foothold.

The configuration of coasts has a great influence upon descents and their prosecution. There are countries where the coasts are steep and present few points of easy access for the ships and the troops to be landed: these few places may be more readily watched, and the descent becomes more difficult.

Finally, there is a strategical consideration connected with descents which may be usefully pointed out. The same principle which forbids a continental army from interposing the mass of its forces between the enemy and the sea requires, on the contrary, that an army landing upon a coast should always keep its principal mass in communication with the shore, which is at once its line of retreat and its base of supplies. For the same reason, its first care should be to make sure of the possession of one fortified harbor/ or at least of a tongue of land which is convenient to a good anchorage and may be easily strengthened by fortifications, in order that in case of reverse the troops may be re-embarked without hurry and loss.

COMMENTARY ON CHAPTER V

*T*his chapter covers operations that lie somewhere between the strategic and the tactical—a scale the 20th century has termed operational. Jomini uses this chapter as a holding pen for topics that do not seem to fit in anywhere else, but which he has included in his book in order to make it comprehensive. Although he seems to slight this chapter by not providing it with an introductory statement, the articles in this chapter are excellent. Article XXXVIII on retreats and pursuits would have been of particular use to a field commander as the military mind tends to shy away from the idea of defeat, and so is least prepared and least practiced in the art of retreat. A successful retreat can mitigate a disaster, while a poorly conducted retreat can turn a minor setback into a calamity. This article at least forces generals to confront failure, giving them responses to minimize the long-term damage of a lost battle.

CHAPTER VI

LOGISTICS; OR, THE PRACTICAL ART OF MOVING ARMIES

ARTICLE XLI
A FEW REMARKS ON LOGISTICS IN GENERAL

*I*s logistics simply a science of detail? Or, on the contrary, is it a general science, forming one of the most essential parts of the art of war? or is it but a term, consecrated by long use, intended to designate collectively the different branches of staff duty,—that is to say, the different means of carrying out in practice the theoretical combinations of the art?

These questions will seem singular to those persons who are firmly convinced that nothing more remains to be said about the art of war, and believe it wrong to search out new definitions where every thing seems already accurately classified. For my own part, I am persuaded that good definitions lead to clear ideas; and I acknowledge some embarrassment in answering these questions which seem so simple.

In the earlier editions of this work I followed the example of other military writers, and called by the name of *logistics* the details of staff duties, which are the subject of regulations for field-service and of special instructions relating to the corps of quartermasters. This was the result of prejudices consecrated by time. The word *logistics* is derived, as we know, from the title of the *major general des logis*, (translated in German by *Quartiermeister,*) an officer whose duty it formerly was to lodge and camp the troops, to give direction to the marches of columns, and to locate them upon the ground. Logistics was then quite limited. But when war began to be waged without camps, movements became more complicated, and the staff officers had more extended functions. The chief of staff began to perform the duty of transmitting the conceptions of the general to the most distant points of the theater of war, and of procuring for him the necessary documents for arranging plans of operations. The chief of staff was called to the assistance of the general in arranging his plans, to give information of them to subordinates in orders and instructions, to explain them and to supervise their execution both in their *ensemble* and in their minute details: his duties were, therefore, evidently connected with all the operations of a campaign.

To be a good chief of staff, it became in this way necessary that a man should be acquainted with all the various branches of the art of war. If the term *logistics* includes all this, the two works of the Archduke Charles, the voluminous treatises of Guibert, Laroche-Aymon, Bousmard, and Ternay, all taken together, would hardly give even an incomplete sketch of what logistics is; for it would be nothing more nor less than the science of applying all possible military knowledge.

It appears from what has been said that the old term *logistics* is insufficient to designate the duties of staff officers, and that the real duties of a corps of such officers, if an attempt be made to instruct them in a proper manner for their performance, should be accurately prescribed by special regulations in accordance with the general principles of the art. Governments should take the precaution to publish well-considered regulations, which should define all the duties of staff officers and should give clear and accurate instructions as to the best methods of performing these duties.

The Austrian staff formerly had such a code of regulations for their government; but it was somewhat behind the times, and was better adapted to the old methods of carrying on war than the present. This is the only work of the kind I have seen. There are, no doubt, others, both public and secret; but I have no knowledge of their existence. Several generals—as, for instance, Grimoard and Thiebaut—have prepared manuals for staff officers, and the new royal corps of France has issued several partial sets of instructions; but there is nowhere to be found a complete manual on the subject.

If it is agreed that the old *logistics* had reference only to details of marches and camps, and, moreover, that the functions of staff officers at the present day are intimately connected with the most important strategical combinations, it must be admitted that logistics includes but a small part of the duties of staff officers; and if we retain the term we must understand it to be greatly extended and developed in signification, so as to embrace not only the duties of ordinary staff officers, but of generals-in-chief.

To convince my readers of this fact, I will mention the principal points that must be included if we wish to embrace in one view every duty and detail relating to the movements of armies and the undertakings resulting from such movements:—

1. The preparation of all the material necessary for setting the army in motion, or, in other words, for opening the campaign. Drawing up orders, instructions, and itineraries for the assemblage of the army and its subsequent launching upon its theater of operations.

2. Drawing up in a proper manner the orders of the general-in-chief for different enterprises, as well as plans of attack in expected battles.

3. Arranging with the chiefs of engineers and artillery the measures to be taken for the security of the posts which are to be used as depots, as well as those to be fortified in order to facilitate the operations of the army.

4. Ordering and directing reconnoissances of every kind, and procuring in this way, and by using spies, as exact information as possible of the positions and movements of the enemy.

5. Taking every precaution for the proper execution of movements ordered by the general. Arranging the march of the different columns, so that all may move in an orderly and connected manner. Ascertaining certainly that the means requisite for the ease and.safety of marches are prepared. Regulating the manner and time of halts.

6. Giving proper composition to advanced guards, rear-guards, flankers, and all detached bodies, and preparing good instructions for their guidance. Providing all the means necessary for the performance of their duties.

7. Prescribing forms and instructions for subordinate commanders or their staff officers, relative to the different methods of drawing up the troops in columns when the enemy is at hand, as well as their formation in the most appropriate manner when the army is to engage in battle, according to the nature of the ground and the character of the enemy.[1]

8. Indicating to advanced guards and other detachments well-chosen points of assembly in case of their attack by superior numbers, and informing them what support they may hope to receive in case of need.

9. Arranging and superintending the march of trains of baggage, munitions, provisions, and ambulances, both with the columns and in their rear, in such manner that they will not interfere with the movements of the troops and will still be near at hand. Taking precautions for order and security, both on the march and when trains are halted and parked.

10. Providing for the successive arrival of convoys of supplies. Collecting all the means of transportation of the country and of the army, and regulating their use.

11. Directing the establishment of camps, and adopting regulations for their safety, good order, and police.

12. Establishing and organizing lines of operations and supplies, as well as lines of communications with these lines for detached bodies. Designating officers capable of organizing and commanding in rear of the army; looking out for the safety of detachments and convoys, furnishing them good instructions, and looking out also for preserving suitable means of communication of the army with its base.

---

1   I refer here to general instructions and forms, which are not to be repeated every day: such repetition would be impracticable.

13. Organizing depots of convalescent, wounded, and sickly men, movable hospitals, and workshops for repairs; providing for their safety.

14. Keeping accurate record of all detachments, either on the flanks or in rear; keeping an eye upon their movements, and looking out for their return to the main column as soon as their service on detachment is no longer necessary; giving them, when required, some center of action, and forming strategic reserves.

15. Organizing marching battalions or companies to gather up isolated men or small detachments moving in either direction between the army and its base of operations.

16. In case of sieges, ordering and supervising the employment of the troops in the trenches, making arrangements with the chiefs of artillery and engineers as to the labors to be performed by those troops and as to their management in sorties and assaults.

17. In retreats, taking precautionary measures for preserving order; posting fresh troops to support and relieve the rear-guard; causing intelligent officers to examine and select positions where the rear-guard may advantageously halt, engage the enemy, check hi pursuit, and thus gain time; making provision in advance for the movement of trains, that nothing shall be left behind, and that they shall proceed in the most perfect order, taking all proper precautions to insure safety.

18. In cantonments, assigning positions to the different corps; indicating to each principal division of the army a place of assembly in case of alarm; taking measures to see that all orders, instructions, and regulations are implicitly observed.

An examination of this long list—which might easily be made much longer by entering into greater detail—will lead every reader to remark that these are the duties rather of the general-in-chief than of staff officers. This truth I announced some time ago; and it is for the very purpose of permitting the general-in-chief to give his whole attention to the supreme direction of the operations that he ought to be provided with staff officers competent to relieve him of details of execution. Their functions are therefore necessarily very intimately connected; and woe to an army where these authorities cease to act in concert! This want of harmony is often seen,—first, because generals are men and have faults, and secondly, because in every army there are found individual interests and pretensions, producing rivalry of the chiefs of staff and hindering them in performing their duties.[1]

It is not to be expected that this treatise shall contain rules for the guidance of staff officers in all the details of their multifarious duties; for, in the first place, every different nation has staff officers with different names and rounds

---

[1]   The chiefs of artillery, of engineers, and of the administrative departments all claim to have direct connection with the general-in-chief, and not with the chief of staff. There should, of course, be no hinderance to the freest intercourse between these high officers and the commander; but he should work with them in presence of the chief of staff, and send him all their correspondence: otherwise, confusion is inevitable.

of duties,—so that I should be obliged to write new rules for each army; in the second place, these details are fully entered into in special books pertaining to these subjects.

I will, therefore, content myself with enlarging a little upon some of the first articles enumerated above:–

1. The measures to be taken by the staff officers for preparing the army to enter upon active operations in the field include all those which are likely to facilitate the success of the first plan of operations. They should, as a matter of course, make sure, by frequent inspections, that the *materiel* of all the arms of the service is in good order: horses, carriages, caissons, teams, harness, shoes, &c. should be carefully examined and any deficiencies supplied. Bridge-trains, engineer-tool trains, *materiel* of artillery, siege-trains if they are to move, ambulances,—in a word, every thing which conies under the head of *materiel*,—should be carefully examined and placed in good order.

If the campaign is to be opened in the neighborhood of great rivers, gun-boats and flying bridges should be prepared, and all the small craft should be collected at the points and at the bank where they will probably be used. Intelligent officers should examine the most favorable points both for embarkations and for landings,—preferring those localities which present the greatest chances of success for a primary establishment on the opposite bank.

The staff officers will prepare all the itineraries that will be necessary for the movement of the several corps of the army to the proper points of assemblage, making every effort to give such direction to the marches that the enemy shall be unable to learn from them any thing relative to the projected enterprise.

If the war is to be offensive, the staff officers arrange with the chief engineer officers what fortifications shall be erected near the base of operations, when *tetes de ponts* or intrenched camps are to be constructed there. If the war is defensive, these works will be built between the first line of defense and the second base.

2. An essential branch of logistics is certainly that which relates to making arrangements of marches and attacks, which are fixed by the general and notice of them given to the proper persons by the chiefs of staff. The next most important qualification of a general, after that of knowing how to form good plans, is, unquestionably, that of facilitating the execution of his orders by their clearness of style. Whatever may be the real business of a chief of staff, the greatness of a commander-in-chief will be always manifested in his plans; but if the general lacks ability the chief of staff should supply it as far as he can, having a proper understanding with the responsible chief.

I have seen two very different methods employed in this branch of the service. The first, which may be styled the old school, consists in issuing daily, for the regulation of the movements of the army, general instructions filled with minute and somewhat pedantic details, so much the more out of place as they are usually addressed to chiefs of corps, who are supposed to be of sufficient experience not to require the same sort of instruction as would be given to junior subalterns just out of school.

The other method is that of the detached orders given by Napoleon to his marshals, prescribing for each one simply what concerned himself, and only informing him what corps were to operate with him, either on the right or the left, but never pointing out the connection of the operations of the whole army.[1] I have good reasons for knowing that he did this designedly, either to surround his operations with an air of mystery, or for fear that more specific orders might fall into the hands of the enemy and assist him in thwarting his plans.

It is certainly of great importance for a general to keep his plans secret; and Frederick the Great was right when he said that if his night-cap knew what was in his head he would throw it into the fire. That kind of secrecy was practicable in Frederick's time, when his whole army was kept closely about him; but when maneuvers of the vastness of Napoleon's are executed, and war is waged as in our day, what concert of action can be expected from generals who are utterly ignorant of what is going on around them?

Of the two systems, the last seems to me preferable. A judicious mean may be adopted between the eccentric conciseness of Napoleon and the minute verbosity which laid down for experienced generals like Barclay, Kleist, and Wittgenstein precise directions for breaking into companies and reforming again in line of battle,—a piece of nonsense all the more ridiculous because the execution of such an order in presence of the enemy is impracticable. It would be sufficient, I think, in such cases, to give the generals special orders relative to their own corps, and to add a few lines in cipher informing them briefly as to the whole plan of the operations and the part they are to take individually in executing it. When a proper cipher is wanting, the order may be transmitted verbally by an officer capable of understanding it and repeating it accurately. Indiscreet revelations need then be no longer feared, and concert of action would be secured.

3. The army being assembled, and being in readiness to undertake some enterprise, the important thing will be to secure as much concert and precision of action as possible, whilst taking all the usual precaution's to gain accurate information of the route it is to pursue and to cover its movements thoroughly.

There are two kinds of marches,—those which are made out of sight of the enemy, and those which are made in his presence, either advancing or retiring. These marches particularly have undergone great changes in late years. Formerly, armies seldom came in collision until they had been several days in presence of each other, and the attacking party had roads opened by pioneers for the columns to move up parallel to each other. At present, the attack is

---

[1]   I believe that at the passage of the Danube before Wagram, and at the opening of the second campaign of 1813, Napoleon deviated from his usual custom by issuing a general order.

made more promptly, and the existing roads usually answer all purposes. It is, however, of importance, when an army is moving, that pioneers and sappers accompany the advanced guard, to increase the number of practicable roads, to remove obstructions, throw small bridges over creeks, &c., if necessary, and secure the means of easy communication between the different corps of the army.

In the present manner of marching, the calculation of times and distances becomes more complicated: the columns having each a different distance to pass over, in determining the hour of their departure and giving them instructions the following particulars must be considered:—1, the distances to be passed over; 2, the amount of *materiel* in each train; 3, the nature of the country; 4, the obstacles placed in the way by the enemy; 5, the fact whether or not it is important for the march to be concealed or open.

Under present circumstances, the surest and simplest method of arranging the movements of the great corps forming the wings of an army, or of all those corps not marching with the column attached to the general head-quarters, will be to trust the details to the experience of the generals commanding those corps,—being careful, however, to let them understand that the most exact punctuality is expected of them. It will then be enough to indicate to them the point to be reached and the object to be attained, the route to be pursued and the hour at which they will be expected to be in position. They should be informed what corps are marching either on the same roads with them or on side-roads to the right or left in order that they may govern themselves accordingly; they should receive whatever news there may be of the enemy, and have a line of retreat indicated to them.[1]

All those details whose object it is to prescribe each day for the chiefs of corps the method of forming their columns and placing them in position are mere pedantry,—more hurtful than useful. To see that they march habitually according to regulation or custom is necessary; but they should be free to arrange their movements so as to arrive at the appointed place and time, at the risk of being removed from their command if they fail to do so without sufficient reason. In retreats, however, which are made along a single road by an army separated into divisions, the hours of departure and halts must be carefully regulated.

Each column should have its own advanced guard and flankers, that its march may be conducted with the usual precautions: it is convenient also, even when they form part of a second line, for the head of each column to be preceded by a few pioneers and sappers, provided with tools for removing obstacles or making repairs in case of accidents; a few of these workmen should also accompany each train: in like manner, a light trestle-bridge train will be found very useful.

---

[1]   Napoleon never did this, because he maintained that no general should ever think seriously of the possibility of being beaten. In many marches it is certainly a useless precaution; but it is often indispensable.

4. The army on the march is often preceded by a general advanced guard, or, as is more frequent in the modern system, the center and each wing may have its special advanced guard. It is customary for the reserves and the center to accompany the head-quarters; and the general advanced guard, when there is one, will usually follow the same road: so that half the army is thus assembled on the central route. Under these circumstances, the greatest care is requisite to prevent obstructing the road. It happens sometimes, however, when the important stroke is to be made in the direction of one of the wings, that the reserves, the general head-quarters, and even the general advanced guard, may be moved in that direction: in this case, all the rules usually regulating the march of the center must be applied to that wing.

Advanced guards should be accompanied by good staff officers, capable of forming correct ideas as to the enemy's movements and of giving an accurate account of them to the general, thus enabling him to make his plans understandingly. The commander of the advanced guard should assist the general in the same way. A general advanced guard should be composed of light troops of all arms, containing some of the *elite* troops of the army as a main body, a few dragoons prepared to fight on foot, some horse-artillery, pontoniers, sappers, &c., with light trestles and pontoons for passing small streams. A few good marksmen will not be out of place. A topographical officer should accompany it, to make a sketch of the country a mile or two on each side of the road. A body of irregular cavalry should always be attached, to spare the regular cavalry and to serve as scouts, because they are best suited to such service.

5. As the army advances and removes farther from its base, it becomes the more necessary to have a good line of operations and of depots which may keep up the connection of the army with its base. The staff officers will divide the depots into departments, the principal depot being established in the town which can lodge and supply the greatest number of men: if there is a fortress suitably situated, it should be selected as the site of the principal depot.

The secondary depots may be separated by distances of from fifteen to thirty miles, usually in the towns of the country. The mean distance apart will be about twenty to twenty-five miles. This will give fifteen depots upon a line of three hundred miles, which should be divided into three or four brigades of depots. Each of these will have a commander and a detachment of troops or of convalescent soldiers, who regulate the arrangements for accommodating troops and give protection to the authorities of the country, (if they remain;) they furnish facilities for transmitting the mails and the necessary escorts; the commander sees that the roads and bridges are kept in good order. If possible, there should be a park of several carriages at each depot, certainly at the principal one in each brigade. The command of all the depots embraced within certain geographical limits should be intrusted to prudent and able general officers; for the security of the communications of the army often depends on their operations.[1] These commands may sometimes become strategic reserves, as was explained in Art. XXIII.; a few good battalions, with the assistance of movable detachments passing continually between the army and the base, will generally be able to keep open the communications.

6. The study of the measures, partly logistical and partly tactical, to be taken by the staff officers in bringing the troops from the order of march to the different orders of battle, is very important, but requires going into such minute detail that I must pass it over nearly in silence, contenting myself with referring my readers to the numerous works specially devoted to this branch of the art of war.

Before leaving this interesting subject, I think a few examples should be given as illustrations of the great importance of a good system of logistics. One of these examples is the wonderful concentration of the French army in the plains of Gera in 1806; another is the entrance of the army upon the campaign of 1815.

In each of these cases Napoleon possessed the ability to make such arrangements that his columns, starting from points widely separated, were concentrated with wonderful precision upon the decisive point of the zone of operations; and in this way he insured the successful issue of the campaign. The choice of the decisive point was the result of a skillful application of the principles of strategy; and the arrangements for moving the troops give us an example of logistics which originated in his own closet. It has been long claimed that Berthier framed those instructions which were conceived with so much precision and usually transmitted with so much clearness; but I have had frequent opportunities of knowing that such was not the truth. The emperor was his own chief staff officer. Provided with a pair of dividers opened to a distance by the scale of from seventeen to twenty miles in a straight line, (which made from twenty-two to twenty-five miles, taking into account the windings of the roads,) bending over and sometimes stretched at full length upon his map, where the positions of his corps and the supposed positions of the enemy were marked by pins of different colors, he was able to give orders for extensive movements with a certainty and precision which were astonishing. Turning his dividers about from point to point on the map, he decided in a moment the number of marches necessary for each of his columns to arrive at the desired point by a certain day; then, placing pins in the new positions, and bearing in mind the rate of marching that he must assign to each column, and the hour of its setting out, he dictated those instructions which are alone enough to make any man famous.

Ney coming from the shores of Lake Constance, Lannes from Upper Swabia, Soult and Davoust from Bavaria and the Palatinate, Bernadotte and Augereau from Franconia, and the Imperial Guard from Paris, were all thus arranged in

---

[1]   It may be objected that in some wars, as where the population is hostile, it may be very difficult, or impracticable, to organize lines of depots. In such cases they will certainly be exposed to great dangers; but these are the very cases where they are most necessary and should be most numerous. The line from Bayonne to Madrid was such a line, which resisted for four years the attacks of the guerrillas,—although convoys were sometimes seized. At one time the line extended as far as Cadiz.

line on three parallel roads, to debouch simultaneously between Saalfeld, Gera, and Plauen, few persons in the army or in Germany having any conception of the object of these movements which seemed so very complicated.

In the same manner, in 1815, when Bluecher had his army quietly in cantonments between the Sambre and the Rhine, and Wellington was attending *fetes* in Brussels, both waiting a signal for the invasion of France, Napoleon, who was supposed to be at Paris entirely engrossed with diplomatic ceremonies, at the head of his guard, which had been but recently reformed in the capital, fell like a thunderbolt upon Charleroi and Bluecher's quarters, his columns arriving from all points of the compass, with rare punctuality, on the 14th of June, in the plains of Beaumont and upon the banks of the Sambre. (Napoleon did not leave Paris until the 12th.)

The combinations described above were the results of wise strategic calculations, but their execution was undoubtedly a masterpiece of logistics. In order to exhibit more clearly the merit of these measures, I will mention, by way of contrast, two cases where faults in logistics came very near leading to fatal consequences. Napoleon having been recalled from Spain in 1809 by the fact of Austria's taking up arms, and being certain that this power intended war, he sent Berthier into Bavaria upon the delicate duty of concentrating the army, which was extended from Braunau as far as Strasbourg and Erfurt. Davoust was returning from the latter city, Oudinot from Frankfort; Massena, who had been on his way to Spain, was retiring toward Ulm by the Strasbourg route; the Saxons, Bavarians, and Wurtembergers were moving from their respective countries. The corps were thus separated by great distances, and the Austrians, who had been long concentrated, might easily break through this spider's web or brush away its threads. Napoleon was justly uneasy, and ordered Berthier to assemble the army at Ratisbon if the war had not actually begun on his arrival, but, if it had, to concentrate it in a more retired position toward Ulm.

The reason for this alternative order was obvious. If the war had begun, Ratisbon was too near the Austrian frontier for a point of assembly, as the corps might thus be thrown separately into the midst of two hundred thousand enemies; but by fixing upon Ulm as the point of rendezvous the army would be concentrated sooner, or, at any rate, the enemy would have five or six marches more to make before reaching-it,—which was a highly-important consideration as the parties were then situated.

No great talent was needed to understand this. Hostilities having commenced, however, but a few days after Berthier's arrival at Munich, this too celebrated chief of staff was so foolish as to adhere to a literal obedience of the order he had received, without conceiving its obvious intention: he not only desired the army to assemble at Ratisbon, but even obliged Davoust to return toward that city, when that marshal had had the good sense to fall back from Amberg toward Ingolstadt.

Napoleon, having, by good fortune, been informed by telegraph of the passage of the Inn twenty-four hours after its occurrence, came with the speed of lightning to Abensberg, just as Davoust was on the point of being surrounded and his army cut in two or scattered by a mass of one hundred and eighty thousand enemies. We know how wonderfully Napoleon succeeded in rallying his army, and what victories he gained on the glorious days of Abensberg, Siegberg, Landshut, Eckmuehl, and Ratisbon, that repaired the faults committed by his chief of staff with his contemptible logistics.

We shall finish these illustrations with a notice of the events which preceded and were simultaneous with the passage of the Danube before the battle of Wagram. The measures taken to bring to a specified point of the island of Lobau the corps of the Viceroy of Italy from Hungary, that of Marmont from Styria, that of Bernadotte from Linz, are less wonderful than the famous imperial decree of thirty-one articles which regulated the details of the passage and the formation of the troops in the plains of Enzersdorf, in presence of one hundred and forty thousand Austrians and five hundred cannon, as if the operation had been a military *fete*. These masses were all assembled upon the island on the evening of the 4th of July; three bridges were immediately thrown over an arm of the Danube one hundred and fifty yards wide, on a very dark night and amidst torrents of rain; one hundred and fifty thousand men passed over the bridges, in presence of a formidable enemy, and were drawn up before mid-day in the plain, three miles in advance of the bridges which they covered by a change of front; the whole being accomplished in less time than might have been supposed necessary had it been a simple maneuver for instruction and after being several times repeated. The enemy had, it is true, determined to offer no serious opposition to the passage; but Napoleon did not know that fact, and the merit of his dispositions is not at all diminished by it.

Singularly enough, however, the chief of staff, although he made ten copies of the famous decree, did not observe that by mistake the bridge of the center had been assigned to Davoust, who had the right wing, whilst the bridge on the right was assigned to Oudinot, who was in the center. These two corps passed each other in the night, and, had it not been for the good sense of the men and their officers, a dreadful scene of confusion might have been the result. Thanks to the supineness of the enemy, the army escaped all disorder, except that arising from a few detachments following corps to which they did not belong. The most remarkable feature of the whole transaction is found in the fact that after such a blunder Berthier should have received the title of Prince of Wagram.

The error doubtless originated with Napoleon while dictating his decree; but should it not have been detected by a chief of staff who made ten copies of the order and whose duty it was to supervise the formation of the troops?

Another no less extraordinary example of the importance of good logistics was afforded at the battle of Leipsic. In fighting this battle, with a defile in rear of the army as at Leipsic, and in the midst of low ground, wooded, and cut up by small streams and gardens, it was highly important to have a number of small bridges, to prepare the banks for approaching them with ease, and to stake out the roads. These precautions would not have prevented the loss of a decisive battle; but they would have saved the lives of a considerable number of men, as well as the guns and carriages that were abandoned on account of the disorder and of there being no roads of escape. The unaccountable blowing up of the bridge of Lindenau was also the result of unpardonable carelessness upon the part of the staff corps, which indeed existed only in name, owing to the manner of Berthier's management of it. We must also agree that Napoleon, who was perfectly conversant with the logistical measures of an offensive campaign, had then never seriously thought what would be proper precautions in the event of defeat, and when the emperor was present himself no one thought of making any arrangement for the future unless by his direction.

To complete what I proposed when I commenced this article, it becomes necessary for me to add some remarks with reference to reconnoissances. They are of two kinds: the first are entirely topographical and statistical, and their object is to gain a knowledge of a country, its accidents of ground, its roads, defiles, bridges, &c., and to learn its resources and means of every kind. At the present day, when the sciences of geography, topography, and statistics are in such an advanced state, these reconnoissances are less necessary than formerly; but they are still very useful, and it is not probable that the statistics of any country will ever be so accurate that they may be entirely dispensed with. There are many excellent books of instruction as to the art of making these reconnoissances, and I must direct the attention of my readers to them.

Reconnoissances of the other kind are ordered when it is necessary to gain information of the movements of the enemy. They are made by detachments of greater or less strength. If the enemy is drawn up in battle-order, the generals-in-chief or the chiefs of staff make the reconnoissance; if he is on the march, whole divisions of cavalry may be thrown out to break through his screen of posts.

ARTICLE XLII

OF RECONNOISSANCES AND OTHER MEANS OF GAINING CORRECT INFORMATION OF THE MOVEMENTS OF THE ENEMY

One of the surest ways of forming good combinations in war would be to order movements only after obtaining perfect information of the enemy's proceedings. In fact, how can any man say what he should do himself, if he is ignorant what his adversary is about? As it is unquestionably of the highest importance to gain this information, so it is a thing of the utmost difficulty, not to say impossibility; and this is one of the chief causes of the great difference between the theory and the practice of war.

From this cause arise the mistakes of those generals who are simply learned men without a natural talent for war, and who have not acquired that practical *coup-d'oeil* which is imparted by long experience in the direction of military operations. It is a very easy matter for a school-man to make a plan for outflanking a wing or threatening a line of communications upon a map, where he can regulate the positions of both parties to suit himself; but when he has opposed to him a skillful, active, and enterprising adversary, whose movements are a perfect riddle, then his difficulties begin, and we see an exhibition of the incapacity of an ordinary general with none of the resources of genius.

I have seen so many proofs of this truth in my long life, that, if I had to put a general to the test, I should have a much higher regard for the man who could form sound conclusions as to the movements of the enemy than for him who could make a grand display of theories,—things so difficult to put in practice, but so easily understood when once exemplified.

There are four means of obtaining information of the enemy's operations. The first is a well-arranged system of espionage; the second consists in reconnoissances made by skillful officers and light troops; the third, in questioning prisoners of war; the fourth, in forming hypotheses of probabilities. This last idea I will enlarge upon farther on. There is also a fifth method,—that of signals. Although this is used rather for indicating the presence of the enemy than for forming conclusions as to his designs, it may be classed with the others.

Spies will enable a general to learn more surely than by any other agency what is going on in the midst of the enemy's camps; for reconnoissances, however well made, can give no information of any thing beyond the line of the advanced guard. I do not mean to say that they should not be resorted to, for we must use every means of gaining information; but I do say that their results are small and not to be depended upon. Reports of prisoners are often useful, but it is generally dangerous to credit them. A skillful chief of staff will always be able to select intelligent officers who can so frame their questions as to elicit important information from prisoners and deserters.

The partisans who are sent to hang around the enemy's lines of operations may doubtless learn something of his movements; but it is almost impossible to communicate with them and receive the information they possess. An extensive system of espionage will generally be successful: it is, however, difficult for a spy to penetrate to the general's closet and learn the secret plans he may form: it is best for him, therefore, to limit himself to information of what he sees with his own eyes or hears from reliable persons. Even when the general receives from his spies information of movements, he still knows nothing of those which may since have taken place, nor of what the enemy is going finally to attempt. Suppose, for example, he learns that such a corps has passed through Jena toward Weimar, and that another has passed through Gera toward Naumburg: he must still ask himself the questions, Where are they going, and what enterprise are they engaged in? These things the most skillful spy cannot learn.

When armies camped in tents and in a single mass, information of the enemy's operations was certain, because reconnoitering-parties could be thrown forward in sight of the camps, and the spies could report accurately their movements; but with the existing organization into corps d'armee which either canton or bivouac, it is very difficult to learn any thing about them. Spies may, however, be very useful when the hostile army is commanded by a great captain or a great sovereign who always moves with the mass of his troops or with the reserves. Such, for example, were the Emperors Alexander and Napoleon. If it was known when they moved and what route they followed, it was not difficult to conclude what project was in view, and the details of the movements of smaller bodies needed not to be attended to particularly.

A skillful general may supply the defects of the other methods by making reasonable and well-founded hypotheses. I can with great satisfaction say that this means hardly ever failed me. Though fortune never placed me at the head of an army, I have been chief of staff to nearly a hundred thousand men, and have been many times called into the councils of the greatest sovereigns of the day, when the question under consideration was the proper direction to give to the combined armies of Europe; and I was never more than two or three times mistaken in my hypotheses and in my manner of solving the difficulties they offered. As I have said before, I have constantly noticed that, as an army can operate only upon the center or one extremity of its front of operations, there are seldom more than three or four suppositions that can possibly be made. A mind fully convinced of these truths and conversant with the principles of war will always be able to form a plan which will provide in advance for the probable contingencies of the future. I will cite a few examples which have come under my own observation.

In 1806, when people in France were still uncertain as to the war with Prussia, I wrote a memoir upon the probabilities of the war and the operations which would take place.

I made the three following hypotheses:—1st. The Prussians will await Napoleon's attack behind the Elbe, and will fight on the defensive as far as the Oder, in expectation of aid from Russia and Austria; 2d. Or they will advance upon the Saale, resting their left upon the frontier of Bohemia and defending the passes of the mountains of Franconia; 3d. Or else, expecting the French by the great Mayence road, they will advance imprudently to Erfurt.

I do not believe any other suppositions could be made, unless the Prussians were thought to be so foolish as to divide their forces, already inferior to the French, upon the two directions of Wesel and Mayence,—a useless mistake, since there had not been a French soldier on the first of these roads since the Seven Years' War.

These hypotheses having been made as above stated, if any one should ask what course Napoleon ought to pursue, it was easy to reply "that the mass of the French army being already assembled in Bavaria, it should be thrown upon the left of the Prussians by way of Grera and Hof, for the gordian knot of the campaign was in that direction, no matter what plan they should adopt."

If they advanced to Erfurt, he could move to Gera, cut their line of retreat, and press them back along the Lower Elbe to the North Sea. If they rested upon the Saale, he could attack their left by way of Hof and Gera, defeat them partially, and reach Berlin before them by way of Leipsic. If they stood fast behind the Elbe, he must still attack them by way of Gera and Hof.

Since Napoleon's direction of operations was so clearly fixed, what mattered it to him to know the details of their movements? Being certain of the correctness of these principles, I did not hesitate to announce, *a month before the war*, that Napoleon would attempt just what he did, and that if the Prussians passed the Saale battles would take place at Jena and Naumburg!

I relate this circumstance not from a feeling of vanity, for if that were my motive I might mention many more of a similar character. I have only been anxious to show that in war a plan of operations may be often arranged, simply based upon the general principles of the art, without much attention being of necessity given to the details of the enemy's movements.

Returning to our subject, I must state that the use of spies has been neglected to a remarkable degree in many modern armies. In 1813 the staff of Prince Schwarzenberg had not a single sou for expenditure for such services, and the Emperor Alexander was obliged to furnish the staff officers with funds from his own private purse to enable them to send agents into Lusatia for the purpose of finding out Napoleon's whereabouts. General Mack at Ulm, and the Duke of Brunswick in 1806, were no better informed; and the French generals in Spain often suffered severely, because it was impossible to obtain spies and to get information as to what was going on around them.

The Russian army is better provided than any other for gathering information, by the use of roving bodies of Cossacks; and history confirms my assertion.

The expedition of Prince Koudacheff, who was sent after the battle of Dresden to the Prince of Sweden, and who crossed the Elbe by swimming and marched in the midst of the French columns as far, nearly, as Wittenberg, is a remarkable instance of this class. The information furnished by the partisan troops of Generals Czernicheff, Benkendorf, Davidoff, and Seslawin was exceedingly valuable. We may recollect it was through a dispatch from Napoleon to the Empress Maria Louisa, intercepted near Chalons by the Cossacks, that the allies were informed of the plan he had formed of falling upon their communications with his whole disposable force, basing his operations upon the fortified towns of Lorraine and Alsace. This highly-important piece of information decided Bluecher and Schwarzenberg to effect a junction of their armies, which the plainest principles of strategy had never previously brought to act in concert except at Leipsic and Brienne.

We know, also, that the warning given by Seslawin to General Doctoroff saved him from being crushed at Borovsk by Napoleon, who had just left Moscow in retreat with his whole army. Doctoroff did not at first credit this news,—which so irritated Seslawin that he effected the capture of a French officer and several soldiers of the guard from the French bivouacs and sent them as proofs of its correctness. This warning, which decided the march of Koutousoff to Maloi-Yaroslavitz, prevented Napoleon from taking the way by Kalouga, where he would have found greater facilities for refitting his army and would have escaped the disastrous days of Krasnoi and the Beresina. The catastrophe which befell him would thus have been lessened, though not entirely prevented.

Such examples, rare as they are, give us an excellent idea of what good partisan troops can accomplish when led by good officers.

I will conclude this article with the following summary:–

1. A general should neglect no means of gaining information of the enemy's movements, and, for this purpose, should make use of reconnoissances, spies, bodies of light troops commanded by capable officers, signals, and questioning deserters and prisoners.

2. By multiplying the means of obtaining information; for, no matter how imperfect and contradictory they may be, the truth may often be sifted from them.

3. Perfect reliance should be placed on none of these means.

4. As it is impossible to obtain exact information by the methods mentioned, a general should never move without arranging several courses of action for himself, based upon probable hypotheses that the relative situation of the armies enables him to make, and never losing sight of the principles of the art.

I can assure a general that, with such precautions, nothing very unexpected can befall him and cause his ruin,—as has so often happened to others; for, unless he is totally unfit to command an army, he should at least be able to form reasonable suppositions as to what the enemy is going to do, and fix for himself a certain line of conduct to suit each of these hypotheses.[1] It cannot be too much insisted upon that the real secret of military genius consists in the ability to make these reasonable suppositions in any case; and, although their number is always small, it is wonderful how much this highly-useful means of regulating one's conduct is neglected.

---

[1] I shall be accused, I suppose, of saying that no event in war can ever occur which may not be foreseen and provided for. To prove the falsity of this accusation, it is sufficient for me to cite the surprises of Cremona, Berg-op-zoom, and Hochkirch. I am still of the opinion, however, that such events even as these might always have been anticipated, entirely or in part, as at least within the limits of probability or possibility.

In order to make this article complete, I must state what is to be gained by using a system of signals. Of these there are several kinds. Telegraphic signals may be mentioned as the most important of all. Napoleon owes his astonishing success at Ratisbon, in 1809, to the fact of his having established a telegraphic communication between the head-quarters of the army and France. He was still at Paris when the Austrian army crossed the Inn at Braunau with the intention of invading Bavaria and breaking through his line of cantonments. Informed, in twenty-four hours, of what was passing at a distance of seven hundred miles, he threw himself into his traveling-carriage, and a week later he had gained two victories under the walls of Ratisbon. Without the telegraph, the campaign would have been lost. This single fact is sufficient to impress us with an idea of its value.

It has been proposed to use portable telegraphs. Such a telegraphic arrangement, operated by men on horseback posted on high ground, could communicate the orders of the center to the extremities of a line of battle, as well as the reports of the wings to the head-quarters. Repeated trials of it were made in Russia; but the project was given up,—for what reason, however, I have not been able to learn. These communications could only be very brief, and in misty weather the method could not be depended upon. A vocabulary for such purposes could be reduced to a few short phrases, which might easily be represented by signs. I think it a method by no means useless, even if it should be necessary to send duplicates of the orders by officers capable of transmitting them with accuracy. There would certainly be a gain of rapidity.[1] attempt of another kind was made in 1794, at the battle of Fleurus, where General Jourdan made use of the services of a balloonist to observe and give notice of the movements of the Austrians. I am not aware that he found the method a very useful one, as it was not again used; but it was claimed at the time that it assisted in gaining him the victory: of this, however, I have great doubts.

It is probable that the difficulty of having a balloonist in readiness to make an ascension at the proper moment, and of his making careful observations upon what is going on below, whilst floating at the mercy of the winds above, has led to the abandonment of this method of gaining information. By giving the balloon no great elevation, sending up with it an officer capable of forming correct opinions as to the enemy's movements, and perfecting a system of signals to be used in connection with the balloon, considerable advantages might be expected from its use. Sometimes the smoke of the battle, and the difficulty of distinguishing the columns, that look like liliputians, so as to know to which party they belong, will make the reports of the balloonists very unreliable. For example, a balloonist would have been greatly embarrassed in deciding, at the battle of Waterloo, whether it was Grouchy or Bluecher who was seen coming up by the Saint-Lambert road; but this uncertainty need not exist where the armies are not so much mixed. I had ocular proof of the advantage to be derived from such observations when I was stationed in the

---

[1]   When the above was written, the magnetic telegraph was not known.—Translators.

spire of Gautsch, at the battle of Leipsic; and Prince Schwarzenberg's aid-de-camp, whom I had conducted to the same point, could not deny that it was at my solicitation the prince was prevailed upon to emerge from the marsh between the Pleisse and the Elster. An observer is doubtless more at his ease in a clock-tower than in a frail basket floating in mid-air; but steeples are not always at hand in the vicinity of battle-fields, and they cannot be transported at pleasure.

There is still another method of signaling, by the use of large fires kindled upon elevated points of the country. Before the invention of the telegraph, they afforded the means of transmitting the news of an invasion from one end of the country to the other. The Swiss have made use of them to call the militia to arms. They have been also used to give the alarm to winter quarters and to assemble the troops more rapidly. The signal-fires may be made still more useful if arranged so as to indicate to the corps of the army the direction of the enemy's threatening movements and the point where they should concentrate to meet him. These signals may also serve on sea-coasts to give notice of descents.

Finally, there is a kind of signals given to troops during an action, by means of military instruments. This method of signals has been brought to greater perfection in the Russian army than in any other I know of. While I am aware of the great importance of discovering a sure method of setting in motion simultaneously a large mass of troops at the will of the commander, I am convinced that it must be a long time before the problem is solved. Signals with instruments are of little use except for skirmishers. A movement of a long line of troops may be made nearly simultaneous by means of a shout begun at one point and passed rapidly from man to man; but these shouts seem generally to be a sort of inspiration, and are seldom the result of an order. I have seen but two cases of it in thirteen campaigns.

COMMENTARY ON CHAPTER VI

*W*hen Jomini writes about logistics, as he does in this chapter, he seems to mean all the planning duties of a staff officer. Arranging marches, drawing up orders and itineraries, directing reconnaissance, coordinating subordinate units, and supplying an army all fall under logistics as Jomini defines the term. This chapter is short, with only two articles, but provides an excellent overview of what was expected of a early 19th century staff officer. After 1870, the almost universal emulation of the more efficient Prussian General Staff system made this model obsolete. Since Jomini spent most of his career in staff positions, this chapter is the one with which he would have had the most personal experience.

CHAPTER VII

OF THE FORMATION OF TROOPS FOR BATTLE, AND THE SEPARATE OR COMBINED USE OF THE THREE ARMS

ARTICLE XLIII
POSTING TROOPS IN LINE OF BATTLE

*H*aving explained in Article XXX. what is to be understood by the term *line of battle*, it is proper to add in what manner it is to be formed, and how the different troops are to be distributed in it.

Before the French Revolution, all the infantry, formed in regiments and brigades, was collected in a single battle-corps, drawn up in two lines, each of which had a right and a left wing. The cavalry was usually placed upon the wings, and the artillery—which at this period was very unwieldy—was distributed along the front of each line. The army camped together, marching by lines or by wings; and, as there were two cavalry wings and two infantry wings, if the march was by wings four columns were thus formed. When they marched by lines, (which was specially applicable to flank movements,) two columns were formed, unless, on account of local circumstances, the cavalry or a part of the infantry had camped in a third line,—which was rare.

This method simplified logistics very much, since it was only necessary to give such orders as the following:—"The army will move in such direction, by lines or by wings, by the right or by the left." This monotonous but simple formation was seldom deviated from; and no better could have been devised as war was carried on in those days.

The French attempted something new at Minden, by forming as many columns as brigades, and opening roads to bring them to the front in line,—a simple impossibility.

If the labor of staff officers was diminished by this method of camping and marching by lines, it must be evident that if such a system were applied to an army of one hundred thousand or one hundred and fifty thousand men, there would be no end to the columns, and the result would be the frequent occurrence of routs like that of Rossbach.

The French Revolution introduced the system of divisions, which broke up the excessive compactness of the old formation, and brought upon the field fractions capable of independent movement on any kind of ground. This change was a real improvement,—although they went from one extreme to the other, by returning nearly to the legionary formation of the Romans. These divisions, composed usually of infantry, artillery, and cavalry, maneuvered and fought separately. They were very much extended, either to enable them to subsist without the use of depots, or with an absurd expectation of prolonging the line in order to outflank that of the enemy. The seven or eight divisions of an army were sometimes seen marching on the same number of roads, ten or twelve miles distant from each other; the head-quarters was at the center, with no other support than five or six small regiments of cavalry of three hundred or four hundred men each, so that if the enemy concentrated the mass of his forces against one of these divisions and beat it, the line was pierced, and the general-in-chief, having no disposable infantry reserve, could do nothing but order a retreat to rally his scattered columns.

Bonaparte in his first Italian campaign remedied this difficulty, partly by the mobility of his army and the rapidity of his maneuvers, and partly by concentrating the mass of his divisions upon the point where the decisive blow was to fall. When he became the head of the government, and saw the sphere of his means and his plans constantly increasing in magnitude, he readily perceived that a stronger organization was necessary: he avoided the extremes of the old system and the new, while still retaining the advantages of the divisional system. Beginning with the campaign of 1800, he organized corps of two or three divisions, which he placed under the command of lieutenant-generals, and formed of them the wings, the center, and the reserve of his army.[1]

This system was finally developed fully at the camp of Boulogne, where he organized permanent army corps under the command of marshals, who had under their orders three divisions of infantry, one of light cavalry, from thirty-six to forty pieces of cannon, and a number of sappers. Each corps was thus a small army, able at need to act independently as an army. The heavy cavalry was collected in a single strong reserve, composed of two divisions of cuirassiers, four of dragoons, and one of light cavalry. The grenadiers and the guard formed an admirable infantry reserve. At a later period—1812—the cavalry was also organized into corps of three divisions, to give greater unity of action to the constantly-increasing masses of this arm. This organization was

---

[1]   Thus, the army of the Rhine was composed of a right wing of three divisions under Lecourbe, of a center of three divisions under Saint-Cyr, and of a left of two divisions under Saint-Suzanne, the general-in-chief having three divisions more as a reserve under his own immediate orders.

as near perfection as possible; and the grand army, that brought about such great results, was the model which all the armies of Europe soon imitated.

Some military men, in their attempts to perfect the art, have recommended that the infantry division, which sometimes has to act independently, should contain three instead of two brigades, because this number will allow one for the center and each wing. This would certainly be an improvement; for if the division contains but two brigades there is an open space left in the center between the brigades on the wings: these brigades, having no common central support, cannot with safety act independently of each other. Besides this, with three brigades in a division, two may be engaged while the third is held in reserve,—a manifest advantage. But, if thirty brigades formed in ten divisions of three brigades are better than when formed in fifteen divisions of two brigades, it becomes necessary, in order to obtain this perfect divisional organization, to increase the numbers of the infantry by one-third, or to reduce the divisions of the army-corps from three to two,—which last would be a serious disadvantage, because the army-corps is much more frequently called upon to act independently than a division, and the subdivision into three parts is specially best for that.[1]

What is the best organization to be given an army just setting out upon a campaign will for a long time to come be a problem in logistics; because it is extremely difficult to maintain the original organization in the midst of the operations of war, and detachments must be sent out continually.

The history of the grand army of Boulogne, whose organization seemed to leave nothing farther to be desired, proves the assertion just made. The center under Soult, the right under Davoust, the left under Ney, and the reserve under Lannes, formed together a regular and formidable battle-corps of thirteen divisions of infantry, without counting those of the guard and the grenadiers. Besides these, the corps of Bernadotte and Marmont detached to the right, and that of Augereau to the left, were ready for action on the flanks. But after the passage of the Danube at Donauwerth every thing was changed. Ney, at first reinforced to five divisions, was reduced to two; the battle-corps was divided partly to the right and partly to the left, so that this fine arrangement was destroyed.

It will always be difficult to fix upon a stable organization. Events are, however, seldom so complicated as those of 1805; and Moreau's campaign of 1800 proves that the original organization may sometimes be maintained, at least for the mass of the army. With this view, it would seem prudent to organize an army in four parts,—two wings, a center, and a reserve. The composition of these parts may vary with the strength of the army; but in order to retain this organization it becomes necessary to have a certain number of divisions out of the general line in order to furnish the necessary detachments. While

---

1   Thirty brigades formed in fifteen divisions of two brigades each will have only fifteen brigades in the first line, while the same thirty brigades formed in ten divisions of three brigades each may have twenty brigades in the first line and ten in the second. But it then becomes necessary to diminish the number of divisions and to have but two in a corps,—which would be a faulty arrangement, because the corps is much more likely to be called upon for independent action than the division.

these divisions are with the army, they may be attached to that part which is to receive or give the heaviest blows; or they may be employed on the flanks of the main body, or to increase the strength of the reserve. Bach of the four great parts of the army may be a single corps of three or four divisions, or two corps of two divisions each. In this last case there would be seven corps, allowing one for the reserve; but this last corps should contain three divisions, to give a reserve to each wing and to the center.

With seven corps, unless several more are kept out of the general line in order to furnish detachments, it may happen that the extreme corps may be detached, so that each wing might contain but two divisions, and from these a brigade might be occasionally detached to flank the march of the army, leaving but three brigades to a wing. This would be a weak order of battle.

These facts lead me to conclude that an organization of the line of battle in four corps of three divisions of infantry and one of light cavalry, with three or four divisions for detachments, would be more stable than one of seven corps, each of two divisions.

But, as every thing depends upon the strength of the army and of the units of which it is composed, as well as upon the character of the operations in which it may be engaged, the arrangement may be greatly varied. I cannot go into these details, and shall simply exhibit the principal combinations that may result from forming the divisions in two or three brigades and the corps in two or three divisions. I have indicated the formation of two infantry corps in two lines, either one behind the other, or side by side. (See Figures from 17 to 28 inclusive.)

Different Formations of Lines of Battle for Two Corps of Infantry.

Figure 17 - Two Corps deployed, One behind the Other

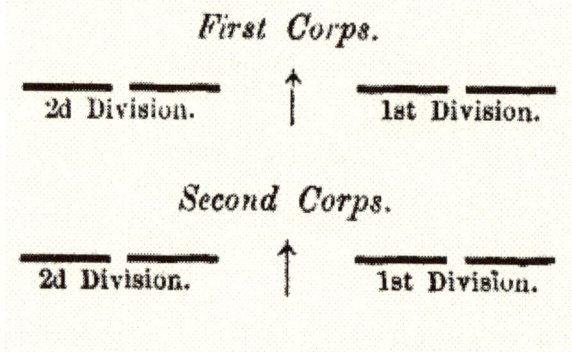

First Corps.

2d Division.   ↑   1st Division.

Second Corps.

2d Division.   ↑   1st Division.

Figure 18 - Two Corps formed Side by Side

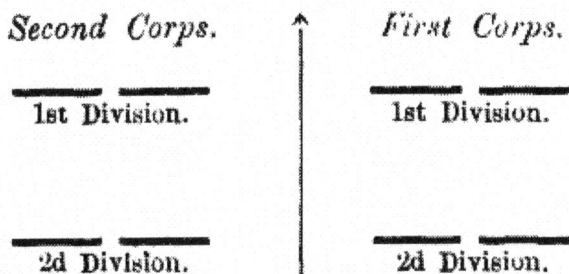

Second Corps. ↑ *First Corps.*

1st Division.   1st Division.

2d Division.   2d Division.

Figure 19 - Two Corps of 2 Divisions of 3 Brigades each

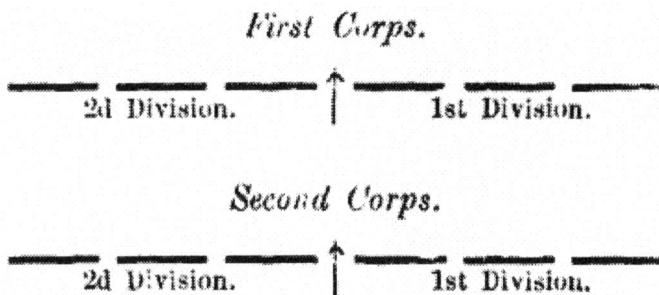

First Corps.

2d Division. ↑ 1st Division.

Second Corps.

2d Division. ↑ 1st Division.

Figure 20 - 2 Corps of 2 Divisions of 3 Brigades each

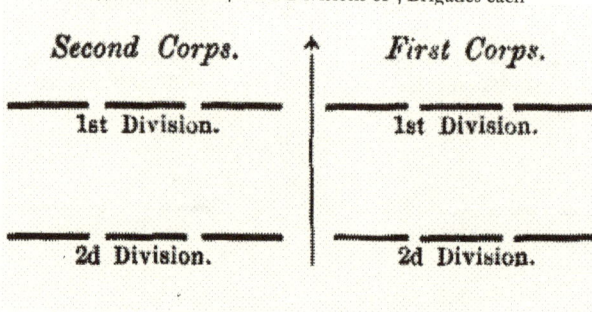

Figure 21 - 2 Corps of 2 Divisions of 3 Brigades each

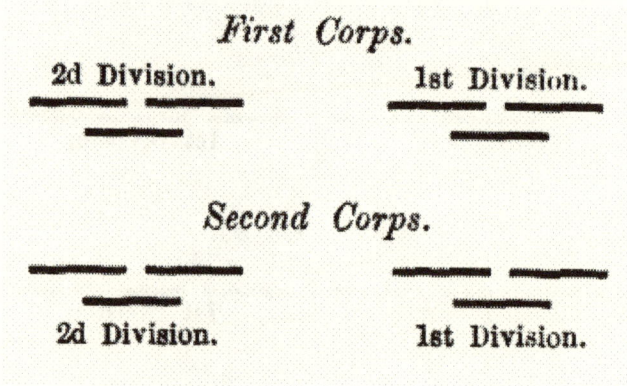

Figure 22

2 Corps of 2 Divisions of 3 Brigades each, placed Side by Side

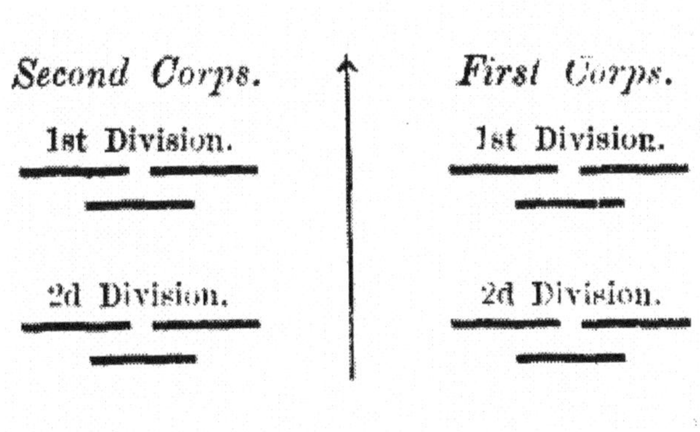

Second Corps.

1st Division.

2d Division.

First Corps.

1st Division.

2d Division.

Formation of Two Corps of Three Divisions of Two Brigades each.

Figure 23

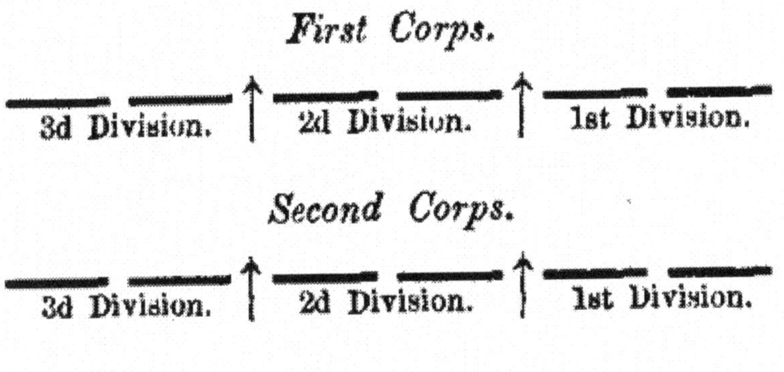

First Corps.

3d Division.   2d Division.   1st Division.

Second Corps.

3d Division.   2d Division.   1st Division.

Figure 24

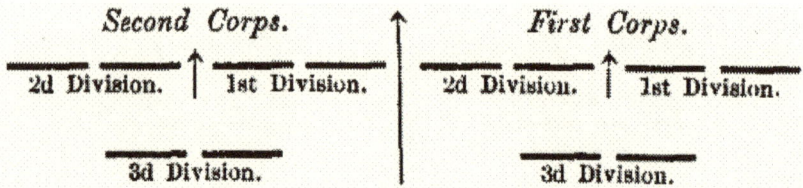

Figure 25

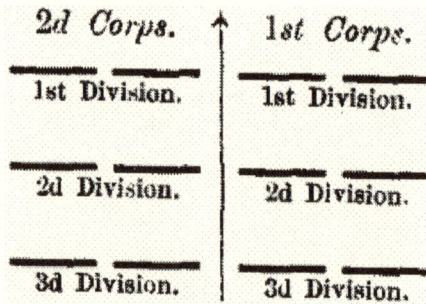

Two Corps of Three Divisions of Three Brigades each.

Figure 26 - Two Divisions in the 1st Line, and one in the 2d Line

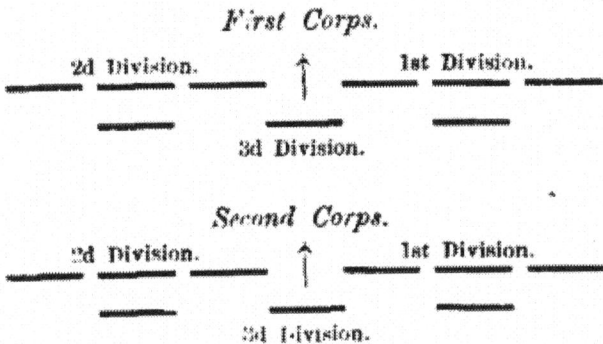

Figure 27 - Same Order with 3d Brigade as Reserve, and the 2 Corps Side by Side

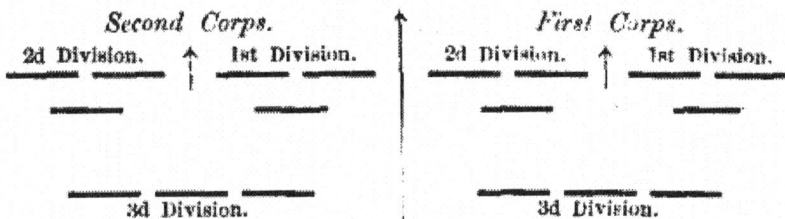

Figure 28

Second Corps.

2d Division. ↑ 1st Division.

3d Division.

First Corps.

2d Division. ↑ 1st Division.

3d Division.

Note.—In all these formations the unit is the brigade in line; but these lines may be formed of deployed battalions, or of battalions in columns of attack by divisions of two companies. The cavalry attached to the corps will be placed on the flanks. The brigades might be so drawn up as to have one regiment in the first line and one in the second.

The question here presents itself, whether it is ever proper to place two corps one behind the other, as Napoleon often did, particularly at Wagram. I think that, except for the reserves, this arrangement may be used only in a position of expectation, and never as an order of battle; for it is much better for each corps to have its own second line and its reserve than to pile up several corps, one behind the other, under different commanders. However much one general may be disposed to support a colleague, he will always object to dividing up his troops for that purpose; and when in the general of the first line he sees not a colleague, but a hated rival, as too frequently happens, it is probable he will be very slow in furnishing the assistance which may be greatly needed. Moreover, a commander whose troops are spread out in a long line cannot execute his maneuvers with near so much facility as if his front was only half as great and was supported by the remainder of his own troops drawn up in rear.

The table below[1] will show that the number of men in an army will have great influence in determining the best formation for it, and that the subject is a complicated one.

---

[1]   Every army has two wings, a center, and a reserve,—in all, four principal subdivisions,—besides accidental detachments.
Below are some of the different formations that may be given to infantry.
1st. In regiments of two battalions of eight hundred men each:–
Div's. Brig's. Batt'ns. Men. Four corps of two divisions each, and three divisions for detachments.... 11 = 22 = 88 = 72,000
Four corps of three divisions each, and three divisions for detachments.... 15 = 30 = 120 = 96,000
Seven corps of two divisions each, and one corps for detachments.... 16 = 32 = 128 = 103,000
2d. In regiments of three battalions, brigades of six battalions:–
Div's. Brig's. Batt'ns. Men. Four corps of two divisions each, besides detachments,..... 11 = 22 = 132 105,000
Four corps of three divisions each, besides detachments.... 15 = 30 = 180 = 144,000
Eight corps of two divisions each.... 16 = 32 = 192 = 154,000

In making our calculations, it is scarcely necessary to provide for the case of such immense masses being in the field as were seen from 1812 to 1815, when a single army contained fourteen corps varying in strength from two to five divisions. With such large numbers nothing better can be proposed than a subdivision into corps of three divisions each. Of these corps, eight would form the main body, and there would remain six for detachments and for strengthening any point of the main line that might require support. If this system be applied to an army of one hundred and fifty thousand men, it would be hardly practicable to employ divisions of two brigades each where Napoleon and the allies used corps.

If nine divisions form the main body,—that is, the wings and the center,— and six others form the reserve and detachments, fifteen divisions would be required, or thirty brigades,—which would make one hundred and eighty battalions, if each regiment contains three battalions. This supposition brings our army up to one hundred and forty-five thousand foot-soldiers and two hundred thousand in all. With regiments of two battalions there would be required one hundred and twenty battalions, or ninety-six thousand infantry; but if each regiment contains but two battalions, each battalion should be one thousand men strong, and this would increase the infantry to one hundred and twenty thousand men and the entire army to one hundred and sixty thousand men. These calculations show that the strength of the minor subdivisions must be carefully considered in arranging into corps and divisions. If an army does not contain more than one hundred thousand men, the formation by divisions is perhaps better than by corps. An example of this was Napoleon's army of 1800.

Having now endeavored to explain the best method of giving a somewhat permanent organization to the main body of an army, it will not be out of place for me to inquire whether this permanency is desirable, and if it is not advantageous to deceive the enemy by frequently changing the composition of corps and their positions.

I admit the advantage of thus deceiving the enemy; but it may be gained while still retaining a quite constant organization of the main body. If the divisions intended for detachments are joined to the wings and the center,— that is, if those parts contain each four divisions instead of three,—and if one or two divisions be occasionally added to the wing which is likely to bear the brunt of an engagement, each wing will be a corps properly of four divisions; but detachments will generally reduce it to three, and sometimes two, while it might, again, be reinforced by a portion of the reserve until it reached five divisions. The enemy would thus never know exactly the strength of the different parts of the line.

If to these numbers we add one-fourth for cavalry, artillery, and engineers, the total force for the above formations may be known.
It is to be observed that regiments of two battalions if eight hundred men each would become very weak at the end of two or three months' campaigning. If they do not consist of three battalions, then each battalion should contain one thousand men.

But I have dwelt sufficiently on these details. It is probable that, whatever be the strength and number of the subdivisions of an army, the organization into corps will long be retained by all the great powers of Europe, and calculations for the arrangement of the line of battle must be made upon that basis.

The distribution of the troops in the line of battle has changed in recent times, as well as the manner of arranging the line. Formerly it was usually composed of two lines, but now of two lines and one or more reserves. In recent[I] conflicts in Europe, when the masses brought into collision were very large, the corps were not only formed in two lines, but one corps was placed behind another, thus making four lines; and, the reserve being drawn up in the same manner, six lines of infantry were often the result, and several of cavalry. Such a formation may answer well enough as a preparatory one, but is by no means the best for battle, as it is entirely too deep.

The classical formation—if I may employ that term—is still two lines for the infantry. The greater or less extent of the battle-field and the strength of an army may necessarily produce greater depth at times; but these cases are the exceptions, because the formation of two lines and the reserves gives sufficient solidity, and enables a greater number of men to be simultaneously engaged.

When an army has a permanent advanced guard, it may be either formed in front of the line of battle or be carried to the rear to strengthen the reserve;[II] but, as has been previously stated, this will not often happen with the present method of forming and moving armies. Each wing has usually its own advanced guard, and the advanced guard of the main or central portion of the army is naturally furnished by the leading corps: upon coming into view of the enemy, these advanced bodies return to their proper positions in line of battle. Often the cavalry reserve is almost entirely with the advanced guard; but this does not prevent its taking, when necessary, the place fixed for it in the line of battle by the character of the position or by the wishes of the commanding general.

From what has been stated above, my readers will gather that very great changes of army organization took place from the time of the revival of the art of war and the invention of gunpowder to the French Revolution, and that to have a proper appreciation of the wars of Louis XIV., of Peter the Great, and of Frederick II., they should consider them from the stand-point of those days.

One portion of the old method may still be employed; and if, by way of example, it may not be regarded as a fundamental rule to post the cavalry on the wings, it may still be a very good arrangement for an army of fifty or sixty thousand men, especially when the ground in the center is not so suitable for the evolutions of cavalry as that near the extremities. It is usual to attach one or two brigades of light cavalry to each infantry corps, those of the center

---

I   The term *recent* here refers to the later wars of Napoleon I.—Translators.
II   As the advanced guard is in presence of the enemy every day, and forms the rear-guard in retreat, it seems but fair at the hour of battle to assign it a position more retired than that in front of the line of battle.

being placed in preference to the rear, whilst those of the wings are placed upon the flanks. If the reserves of cavalry are sufficiently numerous to permit the organization of three corps of this arm, giving one as reserve to the center and one to each wing, the arrangement is certainly a good one. If that is impossible, this reserve may be formed in two columns, one on the right of the left wing and the other on the left of the right wing. These columns may thus readily move to any point of the line that may be threatened.[1]

The artillery of the present day has greater mobility, and may, as formerly, be distributed along the front, that of each division remaining near it. It may be observed, moreover, that, the organization of the artillery having been greatly improved, an advantageous distribution of it may be more readily made; but it is a great mistake to scatter it too much. Few precise rules can be laid down for the proper distribution of artillery. Who, for example, would dare to advise as a rule the filling up of a large gap in a line of battle with one hundred pieces of cannon in a single battery without adequate support, as Napoleon did successfully at Wagram? I do not desire to go here into much detail with reference to the use of this arm, but I will give the following rules:–

1. The horse-artillery should be placed on such ground that it can move freely in every direction.

2. Foot-artillery, on the contrary, and especially that of heavy caliber, will be best posted where protected by ditches or hedges from sudden charges of cavalry. It is hardly necessary for me to add—what every young officer should know already—that too elevated positions are not those to give artillery its greatest effect. Flat or gently-sloping ground is better.

3. The horse-artillery usually maneuvers with the cavalry; but it is well for each army-corps to have its own horse-artillery, to be readily thrown into any desired position. It is, moreover, proper to have horse-artillery in reserve, which may be carried as rapidly as possible to any threatened point. General Benningsen had great cause for self-congratulation at Eylau because he had fifty light guns in reserve; for they had a powerful influence in enabling him to recover himself when his line had been broken through between the center and the left.

4. On the defensive, it is well to place some of the heavy batteries in front, instead of holding them in reserve, since it is desirable to attack the enemy at the greatest possible distance, with a view of checking his forward movement and causing disorder in his columns.

5. On the defensive, it seems also advisable to have the artillery not in reserve distributed at equal intervals in batteries along the whole line, since it is important to repel the enemy at all points. This must not, however, be regarded as an invariable rule; for the character of the position and the designs of the enemy may oblige the mass of the artillery to move to a wing or to the center.

---

1 This disposition of the cavalry, of course, is made upon the supposition that the ground is favorably situated for it. This is the essential condition of every well-arranged line of battle.

6. In the offensive, it is equally advantageous to concentrate a very powerful artillery-fire upon a single point where it is desired to make a decisive stroke, with a view of shattering the enemy's line to such a degree that he will be unable to withstand an attack upon which the fate of the battle is to turn. I shall at another place have more to say as to the employment of artillery in battles.

ARTICLE XLIV

FORMATION AND EMPLOYMENT OF INFANTRY

Infantry is undoubtedly the most important arm of the service, since it forms four-fifths of an army and is used both in the attack and defense of positions. If we must admit that, next to the genius of the general, the infantry arm is the most valuable instrument in gaining a victory, it is no less true that most important aid is given by the cavalry and artillery, and that without their assistance the infantry might at times be very seriously compromised, and at others could achieve only partial success.

We shall not here introduce those old discussions about the shallow and the deep formations, although the question, which was supposed decided, is far from being settled absolutely. The war in Spain and the battle of Waterloo have again given rise to disputes as to the relative advantages of fire and the shallow order, and of columns of attack and the deep order. I will give my own opinion farther on.

There must, however, be no misconception on this subject. The question now is not whether Lloyd was right in wishing to add a fourth rank, armed with pikes, to the infantry formation, with the expectation of producing more effect by the shock when attacking, or opposing a greater resistance when attacked. Every officer of experience knows the difficulty of moving in an orderly manner several deployed battalions in three ranks at close order, and that a fourth rank would increase the disorder without adding any advantage. It is astonishing that Lloyd, who had seen service, should have insisted so much upon the material advantage to be gained by thus increasing the mass of a battalion; for it very rarely happens that such a collision between opposing troops takes place that mere weight decides the contest. If three ranks turn their backs to the enemy, the fourth will not check them. This increase in the number of ranks diminishes the front and the number of men firing upon the defensive, whilst in the offensive there is not near so much mobility as in the ordinary column of attack. It is much more difficult to move eight hundred men in line of battle in four ranks than in three: although in the former case the extent of front is less, the ranks cannot be kept properly closed.

Lloyd's proposal for remedying this diminution of front is so absurd that it is wonderful how a man of talents could have imagined it. He wishes to deploy twenty battalions, and leave between them one hundred and fifty yards, or an interval equal to their front. We may well ask what would befall those battalions thus separated. The cavalry may penetrate the intervals and scatter them like dust before the whirlwind.

But the real question now is, shall the line of battle consist of deployed battalions depending chiefly upon their fire, or of columns of attack, each battalion being formed in column on the central division and depending on its force and impetuosity?

I will now proceed to sum up the particulars bearing upon a decision of the question in hand.

There are, in fact, only five methods of forming troops to attack an enemy:–1, as skirmishers; 2, in deployed lines, either continuous or checkerwise; 3, in lines of battalions formed in column on the central divisions; 4, in deep masses; 5, in small squares.

The skirmishing-order is an accessory; for the duties of skirmishers are, not to form the line of battle, but to cover it by taking advantage of the ground, to protect the movements of columns, to fill up intervals, and to defend the skirts of a position.

These different manners of formation are, therefore, reducible to four: the shallow order, where the line is deployed in three ranks; the half-deep order, formed of a line of battalions in columns doubled on the center or in battalion squares; the mixed order, where regiments are partly in line and partly in column; finally, the deep order, composed of heavy columns of battalions deployed one behind the other.

Figure 29[1]

The formation into two deployed lines with a reserve was formerly used to a great extent: it is particularly suitable on the defensive. These deployed lines may either be continuous, (Fig. 29,) or checkerwise, or in echelons.

Figure 30 - Twelve battalions in columns of attack in two lines, with skirmishers in the intervals

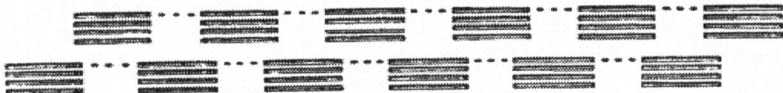

A more compact order is shown in Fig. 30, where each battalion is formed into a column of attack, being by divisions upon the central division. It is really a line of small columns

---

[1]   In this and subsequent figures we suppose a division of twelve battalions.

In the three-rank formation, a battalion with four divisions[I] will have twelve ranks in such a column as shown above: there are in this way too many non-combatants, and the column presents too good a mark for the artillery. To remedy in part these inconveniences, it has been proposed, whenever infantry is employed in columns of attack, to form it in two ranks, to place only three divisions of a battalion one behind the other, and to spread out the fourth as skirmishers in the intervals of the battalions and upon the flanks: when the cavalry charges, these skirmishers may rally behind the other three divisions. (See Fig. 31.) Each battalion would thus have two hundred more men to fire, besides those thrown into the two front ranks from the third. There would be, also, an increase of the whole front. By this arrangement, while having really a depth of but six men, there would be a front of one hundred men, and four hundred men who could discharge their fire-arms, for each battalion. Force and mobility would both be obtained.[II] A battalion of eight hundred men, formed in the ordinary manner in a column of four divisions, has about sixty files in each division, of which the first alone—and only two ranks of that—discharge their pieces. Bach battalion would deliver, therefore, one hundred and twenty shots at a volley, whilst formed in the manner shown in Fig. 31 it would deliver four hundred.

Figure 31

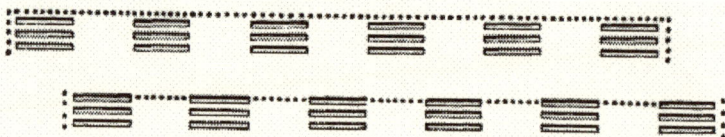

While searching after methods of obtaining more fire when necessary, we must not forget that a column of attack is not intended to fire, and that its fire should be reserved until the last; for if it begins to fire while marching, the whole impulsive effect of its forward movement is lost. Moreover, this shallower order would only be advantageous against infantry, as the column of four divisions in three ranks—forming a kind of solid square—would be better against cavalry. The Archduke Charles found it advantageous at Essling, and particularly at Wagram, to adopt this last order, which was proposed by myself in my chapter on the General Principles of War, published in 1807. The brave cavalry of Bessieres could make no impression upon these small masses.

To give more solidity to the column proposed, the skirmishers might, it is true, be recalled, and the fourth division reformed; but this would be a two-

---

I   The word *division* being used to designate four or five regiments, as well as two companies of a battalion, there is danger of confusion in its use.
II   In the Russian army the skirmishers are taken from the third rank of each division,—which makes the column eight men in depth, instead of twelve, and gives more mobility. To facilitate rallying the skirmishers on the columns, it would be, perhaps, better to take the whole fourth division for that purpose, thus giving nine ranks, or three divisions of three ranks, against infantry, while against cavalry there would be twelve ranks.

rank formation, and would offer much less resistance to a charge than the three-rank formation,—particularly on the flanks. If to remedy this inconvenience it is proposed to form squares, many military men believe that when in two ranks squares would not resist so well as columns. The English squares at Waterloo were, however, only in two ranks, and, notwithstanding the heroic efforts of the French cavalry, only one battalion was broken. I will observe, in conclusion, that, if the two-rank formation be used for the columns of attack, it will be difficult to preserve that in three ranks for deployed lines, as it is scarcely possible to have two methods of formation, or, at any rate, to employ them alternately in the same engagement. It is not probable that any European army, except the English, will undertake to use deployed lines in two ranks. If they do, they should never move except in columns of attack.

I conclude that the system employed by the Russians and Prussians, of forming columns of four divisions in three ranks, of which one may be employed as skirmishers when necessary, is more generally applicable than any other; whilst the other, of which mention has been made, would be suitable only in certain cases and would require a double formation.

Figure 32

There is a mixed order, which was used by Napoleon at the Tagliamento and by the Russians at Eylau, where, in regiments of three battalions, one was deployed to form the first line, and two others to the rear in columns. (See Fig. 32.) This arrangement—which belongs also to the half-deep order—is suitable for the offensive-defensive, because the first line pours a powerful fire upon the enemy, which must throw him into more or less confusion, and the troops formed in columns may debouch through the intervals and fall with advantage upon him while in disorder. This arrangement would probably be improved by placing the leading divisions of the two battalions of the wings upon the same line with the central deployed battalion. There would thus be a half-battalion more to each regiment in the first line,—a by no means unimportant thing for the delivery of fire. There may be reason to fear that, these divisions becoming actively engaged in firing, their battalions which are formed in column to be readily launched against the enemy may not be easily disengaged for that purpose. The order may be useful in many cases. I have therefore indicated it.

Figure 33

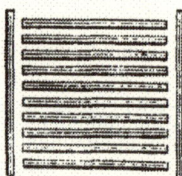

Figure 34

The order in very deep masses (see Figs. 33 and 34) is certainly the most injudicious. In the later wars of Napoleon, twelve battalions were sometimes deployed and closed one upon the other, forming thirty-six ranks closely packed together. Such masses are greatly exposed to the destructive effects of artillery, their mobility and impulsion are diminished, while their strength is not increased. The use of such masses at Waterloo was one cause of the French being defeated. Macdonald's column was more fortunate at Wagram, but at a great sacrifice of life; and it is not probable that this column would have been victorious had it not been for the successes of Davoust and Oudinot on the left of the archduke's line.

When it is decided to risk such a mass, the precaution should certainly be taken of placing on each flank a battalion marching in file, so that if the enemy should charge the mass in flank it need not be arrested in its progress. (See Fig. 33.) Under the protection of these battalions, which may face toward the enemy, the column may continue its march to the point it is expected to reach: otherwise, this large mass, exposed to a powerful converging fire which it has no means of returning, will be thrown into confusion like the column at Fontenoy, or broken as was the Macedonian phalanx by Paulus Emilius.

Squares are good in plains and to oppose an enemy who has a superiority in cavalry. It is agreed that the regimental square is best for the defensive, and the battalion square for the offensive. (See Figs. 35, 36, 37.)

Chapter VII. Of the Formation of Troops for Battle, and
the Separate or Combined Use of the Three Arms

Figure 35 - Division in battalion squares

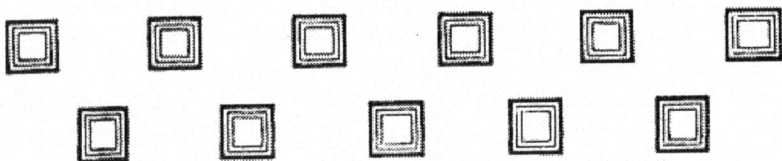

Figure 36 - The same division in long battalion squares

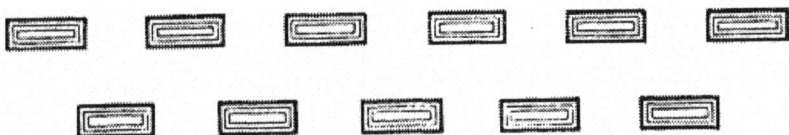

Figure 37 - Squared of regiments of three battalions

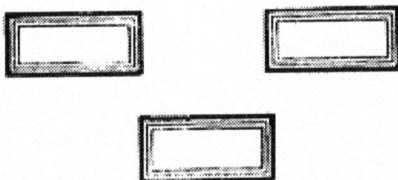

The figures may be perfect squares, or elongated to give a large front and pour a heavier column of fire in the direction of the enemy. A regiment of three battalions will thus form a long square, by wheeling the center battalion half to the right and half to the left.

In the Turkish wars squares were almost exclusively used, because hostilities were carried on in the vast plains of Bessarabia, Moldavia, or Wallachia, and the Turks had an immense force of cavalry. But if the seat of war be the Balkan Mountains or beyond them, and their irregular cavalry be replaced by an army organized according to the proportions usual in Europe, the importance of the square will disappear, and the Russian infantry will show its superiority in Rumelia.

However this may be, the order in squares by regiments or battalions seems suitable for every kind of attack, when the assailant has not the superiority in cavalry and maneuvers on level ground advantageous for the enemy's charges. The elongated square, especially when applied to a battalion of eight companies, three of which would march in front and one on each side, would be much better to make an attack than a deployed battalion. It would not be so good

225

as the column proposed above; but there would be less unsteadiness and more impulsion than if the battalion marched in a deployed line. It would have the advantage, also, of being prepared to resist cavalry.

Squares may also be drawn up in echelons, so as entirely to unmask each other. All the orders of battle may be formed of squares as well as with deployed lines.

It cannot be stated with truth that any one of the formations described is always good or always bad; but there is one rule to the correctness of which every one will assent,—that a formation suitable for the offensive must possess the characteristics of *solidity, mobility,* and *momentum,* whilst for the defensive *solidity* is requisite, and also the power of delivering *as much fire as possible.*

This truth being admitted, it remains yet to be decided whether the bravest troops, formed in columns but unable to fire, can stand long in presence of a deployed line firing twenty thousand musket-balls in one round, and able to fire two hundred thousand or three hundred thousand in five minutes. In the later wars in Europe, positions have often been carried by Russian, French, and Prussian columns with their arms at a shoulder and without firing a shot. This was a triumph of *momentum* and the moral effect it produces; but under the cool and deadly fire of the English infantry the French columns did not succeed so well at Talavera, Busaco, Fuentes-de-Onore, Albuera, and Waterloo.

We must not, however, necessarily conclude from these facts that the advantage is entirely in favor of the shallow formation and firing; for when the French formed their infantry in those dense masses, it is not at all wonderful that the deployed and marching battalions of which they were composed, assailed on all sides by a deadly fire, should have been repulsed. Would the same result have been witnessed if they had used columns of attack formed each of a single battalion doubled on the center? I think not. Before deciding finally as to the superiority of the shallow order, with its facility for firing, over the half-deep order and its momentum, there should be several trials to see how a deployed line would stand an assault from a formation like Fig. 31. These small columns have always succeeded wherever I have seen them tried.

Is it indeed an easy matter to adopt any other order when marching to attack a position? Can an immense deployed line be moved up into action while firing? I think no one will answer affirmatively. Suppose the attempt made to bring up twenty or thirty battalions in line, while firing either by file or by company, to the assault of a well-defended position: it is not very probable they would ever reach the desired point, or, if they did, it would be about as good order as a flock of sheep.

What conclusions shall be drawn from all that has been said? 1. If the deep order is dangerous, the half-deep is excellent for the offensive. 2. The column of attack of single battalions is the best formation for carrying a position by assault; but its depth should be diminished as much as possible, that it may when necessary be able to deliver as heavy a column of fire as possible, and to diminish the effect of the enemy's fire: it ought also to be well covered by skirmishers and supported by cavalry. 3. The formation having the first line deployed and the second in columns is the best-suited to the defensive. 4. Either of them may be successful in the hands of a general of talent, who knows how to use his troops properly in the manner indicated in Articles XVI. and XXX.

Since this chapter was first written, numerous improvements have been made in the arms both of infantry and artillery, making them much more destructive. The effect of this is to incline men to prefer the shallower formations, even in the attack. We cannot, however, forget the lessons of experience; and, notwithstanding the use of rocket-batteries, shrapnel-shot, and the Perkins musket, I cannot imagine a better method of forming infantry for the attack than in columns of battalions. Some persons may perhaps desire to restore to infantry the helmets and breastplates of the fifteenth century, before leading them to the attack in deployed lines. But, if there is a general return to the deployed system, some better arrangement must be devised for marching to the attack than long, continuous lines, and either columns must be used with proper distances for deployment upon arriving near the enemy's position, or lines drawn up checkerwise, or the march must be by the flanks of companies,—all of which maneuvers are hazardous in presence of an enemy who is capable of profiting by the advantages on his side. A skillful commander will use either, or a combination of all, of these arrangements, according to circumstances.

Experience long ago taught me that one of the most difficult tactical problems is that of determining the best formation of troops for battle; but I have also learned that to solve this problem by the use of a single method is an impossibility.

In the first place, the topography of different countries is very various. In some, as Champagne, two hundred thousand men might be maneuvered in deployed lines. In others, as Italy, Switzerland, the valley of the Rhine, half of Hungary, it is barely possible to deploy a division of ten battalions. The degree of instruction of the troops, and their national characteristics, may also have an influence upon the system of formation.

Owing to the thorough discipline of the Russian army and its instruction in maneuvers of every kind, it may maintain in movements in long lines so much order and steadiness as to enable it to adopt a system which would be entirely out of the question for the French or Prussian armies of the present day. My long experience has taught me to believe that nothing is impossible; and I do not belong to the class of men who think that there can be but one type and one system for all armies and all countries.

To approximate as nearly as we can to the solution of the problem, it seems to me, we ought to find out:—1. The best method of moving when in sight of the enemy, but beyond his reach; 2. The best method of coming to close quarters with him; 3. The best defensive order.

In whatever manner we may settle these points, it seems desirable in all cases to exercise the troops—1. In marching in columns of battalions doubled on the center, with a view to deployment, if necessary, when coming into musket-range, or even to attack in column; 2. In marching in continuous deployed lines of eight or ten battalions; 3. In marching in deployed battalions arranged checkerwise,—as these broken lines are more easily moved than continuous lines; 4. In moving to the front by the flanks of companies; 5. In marching to the front in small squares, either in line or checkerwise; 6. In changing front while using these different methods of marching; 7. In changes of front executed by columns of companies at full distance, without deployment,—a more expeditious method than the others of changing front, and the one best suited to all kinds of ground.

Of all the methods of moving to the front, that by the flanks of companies would be the best if it was not somewhat dangerous. In a plain it succeeds admirably, and in broken ground is very convenient. It breaks up a line very much; but by accustoming the officers and privates to it, and by keeping the guides and color-bearers well aligned, all confusion can be avoided. The only objection to it is the danger to which the separated companies are exposed of being ridden down by cavalry. This danger may be avoided by having good cavalry scouts, and not using this formation too near the enemy, but only in getting over the first part of the large interval separating the two armies. At the least sign of the enemy's proximity the line could be reformed instantly, since the companies can come into line at a run. Whatever precautions may be taken, this maneuver should only be practiced with well-disciplined troops, never with militia or raw troops. I have never seen it tried in presence of an enemy,—but frequently at drills, where it has been found to succeed well, especially in changing front.

I have also seen attempts made to march deployed battalions in checkerwise order. They succeeded well; whilst marches of the same battalions in continuous lines did not. The French, particularly, have never been able to march steadily in deployed lines. This checkered order would be dangerous in case of an unexpected charge of cavalry. It may be employed in the first stages of the movement forward, to make it more easy, and the rear battalions would then come into line with the leading ones before reaching the enemy. Moreover, it is easy to form line at the moment of the charge, by leaving a small distance only between the leading and following battalions; for we must not forget that in the checkered order there are not two lines, but a single one, which is broken, to avoid the wavering and disorder observed in the marches of continuous lines.

It is very difficult to determine positively the best formation for making a serious and close attack upon an enemy. Of all the methods I have seen tried, the following seemed to succeed best. Form twenty-four battalions in two lines of battalions in columns doubled on the center ready for deployment: the first line will advance at charging-pace toward the enemy's line to within twice musket-range, and will then deploy at a run; the voltigeur-companies of each battalion will spread out in skirmishing-order, the remaining companies forming line and pouring in a continued fire by file; the second line of columns follows the first, and the battalions composing it pass at charging-step through the intervals of the first line. This maneuver was executed when no enemy was present; but it seems to me an irresistible combination of the advantages of firing and of the column.

Besides these lines of columns, there are three other methods of attacking in the half-deep order.

The first is that of lines composed of deployed battalions with others in column on the wings of those deployed, (Fig. 32). The deployed battalions and the leading divisions of those in column would open fire at half musket-range, and the assault would then be made. The second is that of advancing a deployed line and firing until reaching half musket-range, then throwing forward the columns of the second line through the intervals of the first. The third is the order in echelons, mentioned [earlier and][1] shown in Fig. 15.

Finally, a last method is that of advancing altogether in deployed lines, depending on the superiority of fire alone, until one or the other party takes to its heels,—a case not likely to happen.

I cannot affirm positively which of these methods is the best; for I have not seen them used in actual service. In fact, in real combats of infantry I have never seen any thing but battalions deployed commencing to fire by company, and finally by file, or else columns marching firmly against the enemy, who either retired without awaiting the columns, or repulsed them before an actual collision took place, or themselves moved out to meet the advance. I have seen *melees* of infantry in defiles and in villages, where the heads of columns came in actual bodily collision and thrust each other with the bayonet; but I never saw such a thing on a regular field of battle.

In whatever manner these discussions terminate, they are useful, and should be continued. It would be absurd to discard as useless the fire of infantry, as it would be to give up entirely the half-deep formation; and an army is ruined if forced to adhere to precisely the same style of tactical maneuvers in every country it may enter and against every different nation. It is not so much the mode of formation as the proper combined use of the different arms which will insure victory. I must, however, except very deep masses, as they should be entirely abandoned.

I will conclude this subject by stating that a most vital point to be attended to in leading infantry to the combat is to protect the troops as much as possible from the fire of the enemy's artillery, not by withdrawing them at inopportune moments, but by taking advantage of all inequalities and accidents of the ground to hide them from the view of the enemy. When the assaulting troops have arrived within musket-range, it is useless to calculate upon sheltering

---

1   Publishers change.

them longer: the assault is then to be made. In such cases covers are only suitable for skirmishers and troops on the defensive.

It is generally quite important to defend villages on the front of a position, or to endeavor to take them when held by an enemy who is assailed; but their importance should not be overestimated; for we must never forget the noted battle of Blenheim, where Marlborough and Eugene, seeing the mass of the French infantry shut up in the villages, broke through the center and captured twenty-four battalions which were sacrificed in defending these posts.

For like reasons, it is useful to occupy clumps of trees or brushwood, which may afford cover to the party holding them. They shelter the troops, conceal their movements, cover those of cavalry, and prevent the enemy from maneuvering in their neighborhood. The case of the park of Hougoumont at the battle of Waterloo is a fine example of the influence the possession of such a position, well chosen and strongly defended, may have in deciding the fate of a battle. At Hochkirch and Kolin the possession of the woods was very important.

ARTICLE XLV
CAVALRY

The use a general should make of his cavalry depends, of course, somewhat upon its numerical strength as compared with that of the whole army, and upon its quality. Even cavalry of an inferior character may be so handled as to produce very great results, if set in action at proper moments.

The numerical proportion of cavalry to infantry in armies has varied greatly. It depends on the natural tastes of nations making their people more or less fit for good troopers. The number and quality of horses, also, have something to do with it. In the wars of the Revolution, the French cavalry, although badly organized and greatly inferior to the Austrian, performed wonders. In 1796 I saw what was pompously called the cavalry reserve of the army of the Rhine,—a weak brigade of barely fifteen hundred horses! Ten years later I saw the same reserve consisting of fifteen thousand or twenty thousand horses,—so much had ideas and means changed.

As a general rule, it may be stated that an army in an open country should contain cavalry to the amount of one-sixth its whole strength; in mountainous countries one-tenth will suffice.

The principal value of cavalry is derived from its rapidity and ease of motion. To these characteristics may be added its impetuosity; but we must be careful lest a false application be made of this last.

Whatever may be its importance in the *ensemble* of the operations of war, cavalry can never defend a position without the support of infantry. Its chief duty is to open the way for gaining a victory, or to render it complete by carrying off prisoners and trophies, pursuing the enemy, rapidly succoring a threatened point, overthrowing disordered infantry, covering retreats of infantry and artillery. An army deficient in cavalry rarely obtains a great victory, and finds its retreats extremely difficult.

The proper time and manner of bringing cavalry into action depend upon the ideas of the commander-in-chief, the plan of the battle, the enemy's movements, and a thousand other circumstances which cannot be mentioned here. I can only touch upon the principal things to be considered in its use.

All are agreed that a general attack of cavalry against a line in good order cannot be attempted with much hope of success, unless it be supported by infantry and artillery. At Waterloo the French paid dearly for having violated this rule; and the cavalry of Frederick the Great fared no better at Kunnersdorf. A commander may sometimes feel obliged to push his cavalry forward alone, but generally the best time for charging a line of infantry is when it is already engaged with opposing infantry. The battles of Marengo, Eylau, Borodino, and several others prove this.

There is one case in which cavalry has a very decided superiority over infantry,—when rain or snow dampens the arms of the latter and they cannot fire. Augereau's corps found this out, to their sorrow, at Eylau, and so did the Austrian left at Dresden.

Infantry that has been shaken by a fire of artillery or in any other way may be charged with success. A very remarkable charge of this kind was made by the Prussian cavalry at Hohenfriedberg in 1745. A charge against squares of good infantry in good order cannot succeed.

A general cavalry charge is made to carry batteries of artillery and enable the infantry to take the position more easily; but the infantry must then be at hand to sustain the cavalry, for a charge of this character has only a momentary effect, which must be taken advantage of before the enemy can return offensively upon the broken cavalry. The beautiful charge of the French upon Gosa at the battle of Leipsic, October 16, is a fine example of this kind. Those executed at Waterloo with the same object in view were admirable, but failed because unsupported. The daring charge of Ney's weak cavalry upon Prince Hohenlohe's artillery at Jena is an example of what may be done under such circumstances.

General charges are also made against the enemy's cavalry, to drive it from the field of battle and return more free to act against his infantry.

Cavalry may be successfully thrown against the flank or rear of an enemy's line at the moment of its being attacked in front by the infantry. If repulsed, it may rally upon the army at a gallop, and, if successful, it may cause the loss of the enemy's army. This operation is rarely attempted, but I see no reason why it should not be very good; for a body of cavalry well handled cannot be cut off even if it gets in rear of the enemy. This is a duty for which light cavalry is particularly fitted.

In the defensive, cavalry may also produce very valuable results by opportune dashes at a body of the enemy which has engaged the opposing line and either broken it through or been on the point of doing so. It may regain the advantages lost, change the face of affairs, and cause the destruction of an enemy flushed and disordered by his own success. This was proved at Eylau, where the Russians made a fine charge, and at Waterloo by the English cavalry. The special cavalry of a corps d'armee may charge at opportune moments, either to co-operate in a combined attack, or to take advantage of a false movement of the enemy, or to finish his defeat by pressing him while in retreat.

It is not an easy matter to determine the best mode of attacking, as it depends upon the object in view and other circumstances. There are but four methods of charging,—in columns, in lines at a trot, in lines at a gallop, and in open order,—all of which may be successfully used. In charges in line, the lance is very useful; in *melees*, the saber is much better: hence comes the idea of giving the lance to the front rank, which makes the first onslaught, and the saber to the second rank, which finishes the encounter usually in individual combats. Pistol-firing is of very little use except for outpost-duty, in a charge as foragers, or when light cavalry desires to annoy infantry and draw its fire previous to a charge. I do not know what the carbine is good for; since a body of cavalry armed with it must halt if they wish to fire with any accuracy, and they are then in a favorable condition for the enemy to attack. There are few marksmen who can with any accuracy fire a musket while on horseback and in rapid motion.

I have just said that all the methods of charging may be equally good. It must not be understood, however, that impetuosity always gives the advantage in a shock of cavalry against cavalry: the fast trot, on the contrary, seems to me the best gait for charges in line, because every thing depends, in such a case, upon the *ensemble* and good order of the movement,—things which cannot be obtained in charges at a fast gallop. Galloping is proper against artillery when it is important to get over the ground as rapidly as possible. In like manner, if the cavalry is armed with sabers, it may take the gallop at two hundred yards from the enemy's line if it stands firmly to receive the attack. But if the cavalry is armed with the lance, the fast trot is the proper gait, since the advantageous use of that weapon depends upon the preservation of good order: in a *melee* the lance is almost useless.

If the enemy advances at a fast trot, it does not seem prudent to gallop to meet him; for the galloping party will be much disordered, while the trotting party will not. The only advantage of the gallop is its apparent boldness and the moral effect it produces; but, if this is estimated at its true value by the enemy, it is reasonable to expect his firm and compact mass to be victorious over a body of horsemen galloping in confusion.

In their charges against infantry the Turks and Mamelukes showed the small advantage of mere impetuosity. No cavalry will penetrate where lancers or cuirassiers at a trot cannot. It is only when infantry is much disordered, or their fire poorly maintained, that there is any advantage in the impetuous gallop over the steady trot. To break good squares, cannon and lancers are required, or, better still, cuirassiers armed with lances. For charges in open order there are no better models for imitation than the Turks and the Cossacks.

Whatever method be adopted in charging, one of the best ways of using cavalry is to throw several squadrons opportunely upon the flanks of an enemy's line which is also attacked in front. That this maneuver may be completely successful, especially in charges of cavalry against cavalry, it should be performed at the very moment when the lines come in collision; for a minute too soon or too late its effect may be lost. It is highly important, therefore, that a cavalry commander should have a quick eye, sound judgment, and a cool head.

Much discussion has taken place about the proper manner of arming and organizing cavalry. The lance is the best arm for offensive purposes when a body of horsemen charge in line; for it enables them to strike an enemy who cannot reach them; but it is a very good plan to have a second rank or a reserve armed with sabers, which are more easily handled than the lance in hand-to-hand fighting when the ranks become broken. It would be, perhaps, better still to support a charge of lancers by a detachment of hussars, who can follow up the charge, penetrate the enemy's line, and complete the victory.

The cuirass is the best defensive armor. The lance and the cuirass of strong leather doubled seem to me the best armament for light cavalry, the saber and iron cuirass the best for heavy cavalry. Some military men of experience are inclined even to arm the cuirassiers with lances, believing that such cavalry, resembling very much the men-at-arms of former days, would bear down every thing before them. A lance would certainly suit them better than the musketoon; and I do not see why they should not have lances like those of the light cavalry.

Opinions will be always divided as to those amphibious animals called dragoons. It is certainly an advantage to have several battalions of mounted infantry, who can anticipate an enemy at a defile, defend it in retreat, or scour a wood; but to make cavalry out of foot-soldiers, or a soldier who is equally good on horse or on foot, is very difficult. This might have been supposed settled by the fate of the French dragoons when fighting on foot, had it not been seen that the Turkish cavalry fought quite as well dismounted as mounted. It has been said that the greatest inconvenience resulting from the use of dragoons consists in the fact of being obliged at one moment to make them believe infantry squares cannot resist their charges, and the next moment that a foot-soldier armed with his musket is superior to any horseman in the world. This argument has more plausibility than real force; for, instead of attempting to make men believe such contradictory statements, it would be much more reasonable to tell them that if brave cavalry may break a square, brave foot-soldiers may resist such a charge; that victory does not always depend upon the superiority of the arm, but upon a thousand other things; that the courage of the troops, the presence of mind of the commanders, the opportuneness of maneuvers, the effect of artillery and musketry fire, rain,—mud, even,— have been the causes of repulses or of victories; and, finally, that a brave man, whether on foot or mounted, will always be more than a match for a coward. By impressing these truths upon dragoons, they will believe themselves superior to their adversaries whether they fight on foot or on horseback. This is the case with the Turks and the Circassians, whose cavalry often dismount to fight on foot in a wood or behind a cover, musket in hand, like foot-soldiers.

It requires, however, fine material and fine commanders to bring soldiers to such perfection in knowledge of their duties.

The conviction of what brave men can accomplish, whether on foot or mounted, doubtless induced the Emperor Nicholas to collect the large number of fourteen or fifteen thousand dragoons in a single corps, while he did not consider Napoleon's unfortunate experiment with French dragoons, and was not restrained by the fear of often wanting a regiment of these troops at some particular point. It is probable that this concentration was ordered for the purpose of giving uniformity to the instruction of the men in their duties as foot and mounted soldiers, and that in war they were to be distributed to the different grand divisions of the army. It cannot be denied, however, that great advantages might result to the general who could rapidly move up ten thousand men on horseback to a decisive point and bring them into action as infantry. It thus appears that the methods of concentration and of distribution have their respective advantages and disadvantages. A judicious mean between the extremes would be to attach a strong regiment to each wing of the army and to the advanced guard, (or the rear-guard in a retreat,) and then to unite the remaining troops of this arm in divisions or corps.

Every thing that was said with reference to the formation of infantry is applicable to cavalry, with the following modifications:–

1. Lines deployed checkerwise or in echelons are much better for cavalry than full lines; whilst for infantry lines drawn up checkerwise are too much disconnected, and would be in danger if the cavalry should succeed in penetrating and taking the battalions in flank. The checkerwise formation is only advantageous for infantry in preparatory movements before reaching the enemy, or else for lines of columns which can defend themselves in every direction against cavalry. Whether checkered or full lines be used, the distance between them ought to be such that if one is checked and thrown into confusion the others may not share it. It is well to observe that in the checkered lines the distance may be less than for full lines. In every case the second line should not be full. It should be formed in columns by divisions, or at least there should be left the spaces, if in line, of two squadrons, that may be in column upon the flank of each regiment, to facilitate the passage through of the troops which have been brought up.

2. When the order of columns of attack doubled on the center is used, cavalry should be formed in regiments and infantry only in battalions. The regiments

should contain six squadrons, in order that, by doubling on the center into divisions, three may be formed. If there are only four squadrons, there can be but two lines.

3. The cavalry column of attack should never be formed *en masse* like that of infantry; but there should always be full or half squadron distance, that each may have room to disengage itself and charge separately. This distance will be so great only for those troops engaged. When they are at rest behind the line of battle, they may be closed up, in order to cover less ground and diminish the space to be passed over when brought into action. The masses should, of course, be kept beyond cannon-range.

4. A flank attack being much more to be apprehended by cavalry than in a combat of infantry with infantry, several squadrons should be formed in echelons by platoons on the flanks of a line of cavalry, which may form to the right or left, to meet an enemy coming in that direction.

5. For the same reason, it is important to throw several squadrons against the flanks of a line of cavalry which is attacked in front. Irregular cavalry is quite as good as the regular for this purpose, and it may be better.

6. It is also of importance, especially in cavalry, that the commander-in-chief increase the depth rather than the extent of the formation. For example, in a deployed division of two brigades it would not be a good plan for one brigade to form in a single line behind the other, but each brigade should have one regiment in the first line and one in the second. Each unit of the line will thus have its own proper reserve behind it,—an advantage not to be regarded as trifling; for in a charge events succeed each other so rapidly that it is impossible for a general to control the deployed regiments.

By adopting this arrangement, each general of brigade will be able to dispose of his own reserve; and it would be well, also, to have a general reserve for the whole division. This consideration leads me to think that five regiments would make a good division. The charge may then be made in line by brigades of two regiments, the fifth serving as a general reserve behind the center. Or three regiments may form the line, and two may be in column, one behind each wing. Or it may be preferable to use a mixed order, deploying two regiments and keeping the others in column. This is a good arrangement, because the three regiments, formed in columns by divisions behind the center and flanks of the line, cover those points, and can readily pass the line if it is beaten back. (See Fig. 38.)

Figure 38 - Cavalry division of five regiments. Cavalry deployed should be in checkered order rather than in full lines

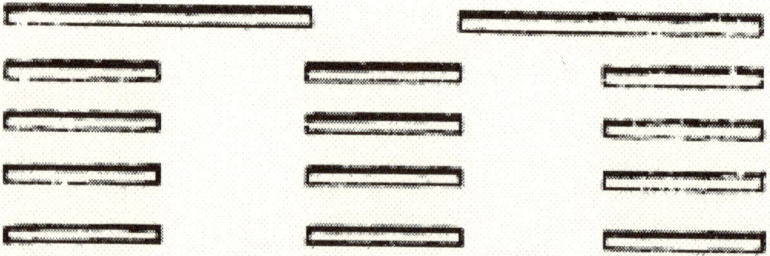

7. Two essential points are regarded as generally settled for all encounters of cavalry against cavalry. One is that the first line must sooner or later be checked; for, even upon the supposition of the first charge being entirely successful, it is always probable that the enemy will bring fresh squadrons to the contest, and the first line must at length be forced to rally behind the second. The other point is that, with troops and commanders on both sides equally good, the victory will remain with the party having the last squadrons in reserve in readiness to be thrown upon the flank of the enemy's line while his front is also engaged.

Attention to these truths will bring us to a just conclusion as to the proper method of forming a large mass of cavalry for battle.

Whatever order be adopted, care must be taken to avoid deploying large cavalry corps in full lines; for a mass thus drawn up is very unmanageable, and if the first line is checked suddenly in its career the second is also, and that without having an opportunity to strike a blow. This has been demonstrated many times. Take as an example the attack made by Nansouty in columns of regiments upon the Prussian cavalry deployed in front of Chateau-Thierry.

In opposing the formation of cavalry in more than two lines, I never intended to exclude the use of several lines checkerwise or in echelons, or of reserves formed in columns. I only meant to say that when cavalry, expecting to make a charge, is drawn up in lines one behind the other, the whole mass will be thrown into confusion as soon as the first line breaks and turns.[1]

With cavalry still more than with infantry the *morale* is very important. The quickness of eye and the coolness of the commander, and the intelligence and bravery of the soldier, whether in the *melee* or in the rally, will oftener be the means of assuring a victory than the adoption of this or that formation. When,

---

1   To disprove my statement, M. Wagner cites the case of the battle of Ramillies, where Marlborough, by a general charge of cavalry in fall lines, succeeded in beating the French drawn up checkerwise. Unless my memory deceives me, the allied cavalry was at first formed checkered in two lines; but the real cause of Marlborough's success was his seeing that Villeroi had paralyzed half his army behind Anderkirch and Gette, and his having the good sense to withdraw thirty-eight squadrons from this wing to reinforce his left, which in this way had twice as many cavalry as the French, and outflanked them. But I cheerfully admit that there may be many exceptions to a rule which I have not laid down more absolutely than all others relating to cavalry tactics,—a tactics, by the way, as changeable as the arm itself.

however, a good formation is adopted and the advantages mentioned above are also present, the victory is more certain; and nothing can excuse the use of a vicious formation.

The history of the wars between 1812 and 1815 has renewed the old disputes upon the question whether regular cavalry will in the end get the better over an irregular cavalry which will avoid all serious encounters, will retreat with the speed of the Parthians and return to the combat with the same rapidity, wearing out the strength of its enemy by continual skirmishing. Lloyd has decided in the negative; and several exploits of the Cossacks when engaged with the excellent French cavalry seem to confirm his opinion. (When I speak of excellent French cavalry, I refer to its impetuous bravery, and not to its perfection; for it does not compare with the Russian or German cavalry either in horsemanship, organization, or in care of the animals.) We must by no means conclude it possible for a body of light cavalry deployed as skirmishers to accomplish as much as the Cossacks or other irregular cavalry. They acquire a habit of moving in an apparently disorderly manner, whilst they are all the time directing their individual efforts toward a common object. The most practiced hussars can never perform such service as the Cossacks, Tscherkesses, and Turks do instinctively.

Experience has shown that irregular charges may cause the defeat of the best cavalry in partial skirmishes; but it has also demonstrated that they are not to be depended upon in regular battles upon which the fate of a war may depend. Such charges are valuable accessories to an attack in line, but alone they can lead to no decisive results.

From the preceding facts we learn that it is always best to give cavalry a regular organization, and furnish them long weapons, not omitting, however, to provide, for skirmishing, &c., an irregular cavalry armed with pistols, lances, and sabers.

Whatever system of organization be adopted, it is certain that a numerous cavalry, whether regular or irregular, must have a great influence in giving a turn to the events of a war. It may excite a feeling of apprehension at distant parts of the enemy's country, it can carry off his convoys, it can encircle his army, make his communications very perilous, and destroy the *ensemble* of his operations. In a word, it produces nearly the same results as a rising *en masse* of a population, causing trouble on the front, flanks, and rear of an army, and reducing a general to a state of entire uncertainty in his calculations.

Any system of organization, therefore, will be a good one which provides for great enlargement of the cavalry in time of war by the incorporation of militia; for they may, with the aid of a few good regular squadrons, be made excellent partisan soldiers. These militia would certainly not possess all the qualities of those warlike wandering tribes who live on horseback and seem born cavalry-soldiers; but they could in a measure supply the places of such. In this respect Russia is much better off than any of her neighbors, both on account of the number and quality of her horsemen of the Don, and the character of the irregular militia she can bring into the field at very short notice.

Twenty years ago I made the following statements in Chapter XXXV. of the Treatise on Grand Military Operations, when writing on this subject:–

"The immense advantages of the Cossacks to the Russian army are not to be estimated. These light troops, which are insignificant in the shock of a great battle, (except for falling upon the flanks,) are terrible in pursuits and in a war of posts. They are a most formidable obstacle to the execution of a general's designs,—because he can never be sure of the arrival and carrying out of his orders, his convoys are always in danger, and his operations uncertain. If an army has had only a few regiments of these half-regular cavalry-soldiers, their real value has not been known; but when their number increases to fifteen thousand or twenty thousand, their usefulness is fully recognized,—especially in a country where the population is not hostile to them.

"When they are in the vicinity, every convoy must be provided with a strong escort, and no movement can be expected to be undisturbed. Much unusual labor is thus made necessary upon the part of the opponent's regular cavalry, which is soon broken down by the unaccustomed fatigue.

"Volunteer hussars or lancers, raised at the time of war breaking out, may be nearly as valuable as the Cossacks, if they are well officered and move freely about from point to point."

In the Hungarians, Transylvanians, and Croats, Austria has resources possessed by few other states. The services rendered by mounted militia have proved, however, that this kind of cavalry may be very useful, if for no other purpose than relieving the regular cavalry of those occasional and extra duties to be performed in all armies, such as forming escorts, acting as orderlies, protecting convoys, serving on outposts, &c. Mixed corps of regular and irregular cavalry may often be more really useful than if they were entirely composed of cavalry of the line,—because the fear of compromising a body of these last often restrains a general from pushing them forward in daring operations where he would not hesitate to risk his irregulars, and he may thus lose excellent opportunities of accomplishing great results.

ARTICLE XLVI

EMPLOYMENT OF ARTILLERY

Artillery is an arm equally formidable both in the offensive and defensive. As an offensive means, a great battery well managed may break an enemy's line, throw it into confusion, and prepare the way for the troops that are to make an assault. As a defensive means, it doubles the strength of a position, not only on account of the material injury it inflicts upon the enemy while at a distance, and the consequent moral effect upon his troops, but also by greatly increasing

the peril of approaching near, and specially within the range of grape. It is
no less important in the attack and defense of fortified places or intrenched
camps; for it is one of the main reliances in modern systems of fortification.

I have already in a former portion of this book given some directions as
to the distribution of artillery in a line of battle; but it is difficult to explain
definitely the proper method of using it in the battle itself. It will not be right
to say that artillery can act independently of the other arms, for it is rather an
accessory. At Wagram, however, Napoleon threw a battery of one hundred
pieces into the gap left by the withdrawal of Massena's corps, and thus held in
check the Austrian center, notwithstanding their vigorous efforts to advance.
This was a special case, and should not be often imitated.

I will content myself with laying down a few fundamental rules, observing
that they refer to the present state of artillery service, (1838.) The recent
discoveries not yet being fully tested, I shall say little with reference to them.

1. In the offensive, a certain portion of the artillery should concentrate its fire
upon the point where a decisive blow is to be struck. Its first use is to shatter
the enemy's line, and then it assists with its fire the attack of the infantry and
cavalry.

2. Several batteries of horse-artillery should follow the offensive movements
of the columns of attack, besides the foot-batteries intended for the same
purpose. Too much foot-artillery should not move with an offensive column.
It may be posted so as to co-operate with the column without accompanying it.
When the cannoneers can mount the boxes, it may have greater mobility and
be advanced farther to the front.

3. It has already been stated that half of the horse-artillery should be held in
reserve, that it may be rapidly moved to any required point.[1] For this purpose
it should be placed upon the most open ground, whence it can move readily
in every direction. I have already indicated the best positions for the heavy
calibers.

4. The batteries, whatever may be their general distribution along the
defensive line, should give their attention particularly to those points where
the enemy would be most likely to approach, either on account of the facility
or the advantage of so doing. The general of artillery should therefore know
the decisive strategic and tactical points of the battle-field, as well as the
topography of the whole space occupied. The distribution of the reserves of
artillery will be regulated by these.

5. Artillery placed on level ground or ground sloping gently to the front is
most favorably situated either for point-blank or ricochet firing: a converging
fire is the best.

6. It should be borne in mind that the chief office of all artillery in battles
is to overwhelm the enemy's troops, and not to reply to their batteries. It is,
nevertheless, often useful to fire at the batteries, in order to attract their fire.

---

[1]   Greater mobility is now given to foot-artillery by mounting the men on the boxes.

A third of the disposable artillery may be assigned this duty, but two-thirds at least should be directed against the infantry and cavalry of the enemy.

7. If the enemy advance in deployed lines, the batteries should endeavor to cross their fire in order to strike the lines obliquely. If guns can be so placed as to enfilade a line of troops, a most powerful effect is produced.

8. When the enemy advance in columns, they may be battered in front. It is advantageous also to attack them obliquely, and especially in flank and reverse. The moral effect of a reverse fire upon a body of troops is inconceivable; and the best soldiers are generally put to flight by it. The fine movement of Ney on Preititz at Bautzen was neutralized by a few pieces of Kleist's artillery, which took his columns in flank, checked them, and decided the marshal to deviate from the excellent direction he was pursuing. A few pieces of light artillery, thrown at all hazards upon the enemy's flank, may produce most important results, far overbalancing the risks run.

9. Batteries should always have supports of infantry or cavalry, and especially on their flanks. Cases may occur where the rule may be deviated from: Wagram is a very remarkable example of this.

10. It is very important that artillerists, when threatened by cavalry, preserve their coolness. They should fire first solid shot, next shells, and then grape, as long as possible. The infantry supports should, in such a case, form squares in the vicinity, to shelter the horses, and, when necessary, the cannoneers. When the infantry is drawn up behind the pieces, large squares of sufficient size to contain whatever they should cover are best; but when the infantry is on the flanks, smaller squares are better. Rocket-batteries may also be very efficient in frightening the horses.

11. When infantry threatens artillery, the latter should continue its fire to the last moment, being careful not to commence firing too soon. The cannoneers can always be sheltered from an infantry attack if the battery is properly supported. This is a case for the co-operation of the three arms; for, if the enemy's infantry is thrown into confusion by the artillery, a combined attack upon it by cavalry and infantry will cause its destruction.

12. The proportions of artillery have varied in different wars. Napoleon conquered Italy in 1800 with forty or fifty pieces,—whilst in 1812 he invaded Russia with one thousand pieces thoroughly equipped, and failed. These facts show that any fixed rule on the subject is inadmissible. Usually three pieces to a thousand combatants are allowed; but this allowance will depend on circumstances.

The relative proportions of heavy and light artillery vary also between wide limits. It is a great mistake to have too much heavy artillery, whose mobility must be much less than that of the lighter calibers. A remarkable proof of the great importance of having a strong artillery-armament was given by Napoleon after the battle of Eylau. The great havoc occasioned among his troops by the numerous guns of the Russians opened his eyes to the necessity of increasing his own. With wonderful vigor, he set all the Prussian arsenals to work, those along the Rhine, and even at Metz, to increase the number of his pieces, and to cast new ones in order to enable him to use the munitions previously captured. In three months he doubled the *materiel* and *personnel* of his artillery, at a distance of one thousand miles from his own frontiers,—a feat without a parallel in the annals of war.

13. One of the surest means of using the artillery to the best advantage is to place in command of it a general who is at once a good strategist and tactician. This chief should be authorized to dispose not only of the reserve artillery, but also of half the pieces attached to the different corps or divisions of the army. He should also consult with the commanding general as to the moment and place of concentration of the mass of his artillery in order to contribute most to a successful issue of the day, and he should never take the responsibility of thus massing his artillery without previous orders from the commanding general.

ARTICLE XLVII

OF THE COMBINED USE OF THE THREE ARMS

To conclude this Summary in a proper manner, I ought to treat of the combined use of the three arms; but I am restrained from so doing by considering the great variety of points necessary to be touched upon if I should attempt to go into an examination of all the detailed operations that would arise in the application of the general rules laid down for each of the arms.

Several authors—chiefly German—have treated this subject very extensively, and their labors are valuable principally because they consist mainly of citations of numerous examples taken from the actual minor engagements of the later wars. These examples must indeed take the place of rules, since experience has shown that fixed rules on the subject cannot be laid down. It seems a waste of breath to say that the commander of a body of troops composed of the three arms should employ them so that they will give mutual support and assistance; but, after all, this is the only fundamental rule that can be established, for the attempt to prescribe for such a commander a special course of conduct in every case that may arise, when these cases may be infinitely varied, would involve him in an inextricable labyrinth of instructions. As the object and limits of this Summary do not allow me to enter upon the consideration of such details, I can only refer my readers to the best works which do treat of them.

I have said all I can properly say when I advise that the different arms be posted in conformity with the character of the ground, according to the object in view and the supposed designs of the enemy, and that they be used simultaneously in the manner best suited to them, care being taken to enable them to afford mutual support. A careful study of the events of previous wars, and especially experience in the operations of war, will give an officer correct ideas on these points, and the ability to use, at the right time and place, his knowledge of the properties of the three arms, either single or combined.

CONCLUSION

I am constrained to recapitulate the principal facts which may be regarded as fundamental in war. War in its *ensemble* is not a science, but an art. Strategy, particularly, may indeed be regulated by fixed laws resembling those of the positive sciences, but this is not true of war viewed as a whole. Among other things, combats may be mentioned as often being quite independent of scientific combinations, and they may become essentially dramatic, personal qualities and inspirations and a thousand other things frequently being the controlling elements. The passions which agitate the masses that are brought into collision, the warlike qualities of these masses, the energy and talent of their commanders, the spirit, more or less martial, of nations and epochs,[1]— in a word, every thing that can be called the poetry and metaphysics of war,— will have a permanent influence on its results.

Shall I be understood as saying that there are no such things as tactical rules, and that no theory of tactics can be useful? What military man of intelligence would be guilty of such an absurdity? Are we to imagine that Eugene and Marlborough triumphed simply by inspiration or by the superior courage and discipline of their battalions? Or do we find in the events of Turin, Blenheim, and Ramillies maneuvers resembling those seen at Talavera, Waterloo, Jena, or Austerlitz, which were the causes of the victory in each case? When the application of a rule and the consequent maneuver have procured victory a hundred times for skillful generals, and always have in their favor the great probability of leading to success, shall their occasional failure be a sufficient reason for entirely denying their value and for distrusting the effect of the study of the art? Shall a theory be pronounced absurd because it has only three-fourths of the whole number of chances of success in its favor?

The *morale* of an army and its chief officers has an influence upon the fate of a war; and this seems to be due to a certain physical effect produced by the moral cause. For example, the impetuous attack upon a hostile line of twenty thousand brave men whose feelings are thoroughly enlisted in their cause will produce a much more powerful effect than the attack of forty thousand demoralized or apathetic men upon the same point.

Strategy, as has already been explained, is the art of bringing the greatest part of the forces of an army upon the important point of the theater of war or of the zone of operations.

---

[1] The well-known Spanish proverb, *He was brave on such a day*, may be applied to nations as to individuals. The French at Rossbach were not the same people as at Jena, nor the Prussians at Prentzlow as at Dennewitz.

Chapter VII. Of the Formation of Troops for Battle, and the Separate or Combined Use of the Three Arms

Tactics is the art of using these masses at the points to which they shall have been conducted by well-arranged marches; that is to say, the art of making them act at the decisive moment and at the decisive point of the field of battle. When troops are thinking more of flight than of fight, they can no longer be termed active masses in the sense in which I use the term.

A general thoroughly instructed in the theory of war, but not possessed of military *coup-d'oeil*, coolness, and skill, may make an excellent strategic plan and be entirely unable to apply the rules of tactics in presence of an enemy: his projects will not be successfully carried out, and his defeat will be probable. If he be a man of character, he will be able to diminish the evil results of his failure, but if he lose his wits he will lose his army.

The same general may, on the other hand, be at once a good tactician and strategist, and have made all the arrangements for gaining a victory that his means will permit: in this case, if he be only moderately seconded by his troops and subordinate officers, he will probably gain a decided victory. If, however, his troops have neither discipline nor courage, and his subordinate officers envy and deceive him,[1] he will undoubtedly see his fine hopes fade away, and his admirable combinations can only have the effect of diminishing the disasters of an almost unavoidable defeat.

No system of tactics can lead to victory when the *morale* of an army is bad; and even when it may be excellent the victory may depend upon some occurrence like the rupture of the bridges over the Danube at Essling. Neither will victories be necessarily gained or lost by rigid adherence to or rejection of this or that manner of forming troops for battle.

These truths need not lead to the conclusion that there can be no sound rules in war, the observance of which, the chances being equal, will lead to success. It is true that theories cannot teach men with mathematical precision what they should do in every possible case; but it is also certain that they will always point out the errors which should be avoided; and this is a highly-important consideration, for these rules thus become, in the hands of skillful generals commanding brave troops, means of almost certain success.

The correctness of this statement cannot be denied; and it only remains to be able to discriminate between good rules and bad. In this ability consists the whole of a man's genius for war. There are, however, leading principles which assist in obtaining this ability. Every maxim relating to war will be good if it indicates the employment of the greatest portion of the means of action at the decisive moment and place. In Chapter III. I have specified all the strategic combinations which lead to such a result. As regards tactics, the principal thing to be attended to is the choice of the most suitable order of battle for the object in view. When we come to consider the action of masses on the field, the means to be used may be an opportune charge of cavalry, a strong battery put in position and unmasked at the proper moment, a column of infantry making

---

[1] The unskillful conduct of a subordinate who is incapable of understanding the merit of a maneuver which has been ordered, and who will commit grave faults in its execution, may produce the same result of causing the failure of the plans of an excellent commander.

a headlong charge, or a deployed division coolly and steadily pouring upon the enemy a fire, or they may consist of tactical maneuvers intended to threaten the enemy's flanks or rear, or any other maneuver calculated to diminish the confidence of the adversary. Each of these things may, in a particular case, be the cause of victory. To define the cases in which each should be preferred is simply impossible.

If a general desires to be a successful actor in the great drama of war, his first duty is to study carefully the theater of operations, that he may see clearly the relative advantages and disadvantages it presents for himself and his enemies. This being done, he can understandingly proceed to prepare his base of operations, then to choose the most suitable zone of operations for his main efforts, and, in doing so, keep constantly before his mind the principles of the art of war relative to lines and fronts of operations. The offensive army should particularly endeavor to cut up the opposing army by skillfully selecting objective points of maneuver; it will then assume, as the objects of its subsequent undertakings, geographical points of more or less importance, depending upon its first successes.

The defensive army, on the contrary, should endeavor, by all means, to neutralize the first forward movement of its adversary, protracting operations as long as possible while not compromising the fate of the war, and deferring a decisive battle until the time when a portion of the enemy's forces are either exhausted by labors, or scattered for the purpose of occupying invaded provinces, masking fortified places, covering sieges, protecting the line of operations, depots, &c.

Up to this point every thing relates to a first plan of operations; but no plan can provide with certainty for that which is uncertain always,—the character and the issue of the first conflict. If your lines of operations have been skillfully chosen and your movements well concealed, and if on the other hand your enemy makes false movements which permit you to fall on fractions of his army, you may be successful in your campaign, without fighting general battles, by the simple use of your strategic advantages. But if the two parties seem about equally matched at the time of conflict, there will result one of those stupendous tragedies like Borodino, Wagram, Waterloo, Bautzen, and Dresden, where the precepts of grand tactics, as indicated in the chapter on that subject, must have a powerful influence.

If a few prejudiced military men, after reading this book and carefully studying the detailed and correct history of the campaigns of the great masters of the art of war, still contend that it has neither principles nor rules, I can only pity them, and reply, in the famous words of Frederick, that "a mule which had made twenty campaigns under Prince Eugene would not be a better tactician than at the beginning."

Correct theories, founded upon right principles, sustained by actual events of wars, and added to accurate military history, will form a true school of instruction for generals. If these means do not produce great men, they will at least produce generals of sufficient skill to take rank next after the natural masters of the art of war.

COMMENTARY ON CHAPTER VII

*T*his chapter, even more than Chapter 4, is primarily of historical interest. Unlike Chapter 3 on strategy, which still has some relevance in the 21st century, Chapter 7's insights are all outdated. On the other hand, when used as a supplementary text to 18th and early 19th century military history, these insights are fascinating. The discussion of the virtues and vices of various corps and divisional organizations is an excellent commentary on the fumbling attempts that Napoleon's ancient regime opponents made to emulate his system of corps, as well as an implicit criticism of Napoleon's lack of a uniform divisional and corps organization. American Civil War historians will recognize in Jomini's organizational comments a critique of the difference between Robert E. Lee's organization of the Army of Northern Virginia into two (later three) large corps and the organization of the Army of the Potomac into seven smaller corps.

The description of various formations a general might employ in his divisions is of great use to anyone struggling to understand the differences between the standard formations used in the Napoleonic age and the benefits that a particular formation would have given its commander. For example, the differences in defensive firepower between the English line formation and the habitual Prussian use of columns accounts for much of the difference in outcomes at the battles of Waterloo and Ligny. On the other hand, maneuverability and flexibility of column formations made it possible for Blucher's Prussians to come to Wellington's aid at Waterloo. This section has less utility for those interested in the Civil War, as the line was used almost exclusively in tactics in the 1860s.

SUPPLEMENT
TO THE
SUMMARY OF THE ART OF WAR

*M*y Summary of the Art of War, published in 1836, to assist in the military instruction of the Hereditary Grand Duke of Russia, contained a concluding article that was never printed. I deem it expedient to give it now in the form of a supplement, and add a special article upon the means of acquiring a certain and ready strategic *coup-d'oeil*.

It is essential for the reader of my Summary to understand clearly that in the military science, as in every other, the study of details is easy for the man who has learned how to seize the fundamental features to which all others are secondary. I am about to attempt a development of these elements of the art; and my readers should endeavor to apprehend them clearly and to apply them properly.

I cannot too often repeat that the theory of the great combinations of war is in itself very simple, and requires nothing more than ordinary intelligence and careful consideration. Notwithstanding its simplicity, many learned military men have difficulty in grasping it thoroughly. Their minds wander off to accessory details, in place of fixing themselves on first causes, and they go a long way in search of what is just within their reach if they only would think so.

Two very different things must exist in a man to make him a general: *he must know how to arrange a good plan of operations, and how to carry it to a successful termination*. The first of these talents may be a natural gift, but it may also be acquired and developed by study. The second depends more on individual character, is rather a personal attribute, and cannot be created by study, although it may be improved.

It is particularly necessary for a monarch or the head of a government to possess the first of these talents, because in such case, although he may not have the ability to execute, he can arrange plans of operations and decide correctly as to the excellence or defects of those submitted to him by others. He is thus enabled to estimate properly the capacity of his generals, and when he finds a general producing a good plan, and having firmness and coolness, such a man may be safely trusted with the command of an army.

If, on the other hand, the head of a state is a man of executive ability, but not possessing the faculty of arranging wise military combinations, he will be likely to commit all the faults that have characterized the campaigns of many celebrated warriors who were only brave soldiers without being at all improved by study.

From the principles which I have laid down, and their application to several famous campaigns, my readers will perceive that the theory of the great combinations of war may be summed up in the following truths.

The science of strategy consists, in the first place, in knowing how to choose well a theater of war and to estimate correctly that of the enemy. To do this, a general must accustom himself to decide as to the importance of decisive points,—which is not a difficult matter when he is aided by the hints I have given on the subject, particularly in Articles from XVIII. to XXII.

The art consists, next, in a proper employment of the troops upon the theater of operations, whether offensive or defensive. (See Article XVII.) This employment of the forces should be regulated by two fundamental principles: the first being, *to obtain by free and rapid movements the advantage of bringing the mass of the troops against fractions of the enemy; the second, to strike in the most decisive direction,*—that is to say, in that direction where the consequences of his defeat may be most disastrous to the enemy, while at the same time his success would yield him no great advantages.

The whole science of great military combination is comprised in these two fundamental truths. Therefore, all movements that are disconnected or more extended than those of the enemy would be grave faults; so also would the occupation of a position that was too much cut up, or sending out a large detachment unnecessarily. On the contrary, every well-connected, compact system of operations would be wise; so also with central strategic lines, and every strategic position less extended than the enemy's.

The application of these fundamental principles is also very simple. If you have one hundred battalions against an equal number of the enemy's, you may, by their mobility and by taking the initiative, bring eighty of them to the decisive point while employing the remaining twenty to observe and deceive half of the opposing army. You will thus have eighty battalions against fifty at the point where the important contest is to take place. You will reach this point by rapid marches, by interior lines, or by a general movement toward one extremity of the hostile line. I have indicated the cases in which one or the other of these means is to be preferred.

In arranging a plan of operations, it is important to remember *"that a strategic theater, as well as every position occupied by an army, has a center and two extremities."* A theater has usually three zones,—a right, a left, and a central.

In choosing a zone of operations, select one,—1, that will furnish a safe and advantageous base; 2, in which the least risk will be run by yourself, while the enemy will be most exposed to injury; 3, bearing in mind the antecedent situations of the two parties, and, 4, the dispositions and inclinations of the powers whose territories are near the theater of war.

One of the zones will always be decidedly bad or dangerous, while the other two will be more or less suitable according to circumstances.

The zone and base being fixed upon, the object of the first attempts must be selected. This is choosing an objective of operations. There are two very different kinds: some, that are called *territorial or geographical objectives*, refer simply to an enemy's line of defense which it is desired to get possession of, or a fortress or intrenched camp to be captured; *the others, on the contrary, consist entirely in the destruction or disorganization of the enemy's forces, without giving attention to geographical points of any kind.* This was the favorite objective of Napoleon.[1]

I can profitably add nothing to what I have already written on this point, *and, as the choice of the objective is by far the most important thing in a plan of operations,* I recommend the whole of Article XIX.

The objective being determined upon, the army will move toward it by one or two lines of operations, care being taken to conform to the fundamental principle laid down, and to avoid double lines, unless the character of the theater of war makes it necessary to use them, or the enemy is very inferior either in the number or the quality of his troops. Article XXI. treats this subject fully. If two geographical lines are used, it is essential to move the great mass of the forces along the most important of them, and to occupy the secondary line by detachments having a concentric direction, if possible, with the main body.

The army, being on its way toward the objective, before arriving in presence of the enemy and giving battle, occupies daily or temporary strategic positions: the front it embraces, or that upon which the enemy may attack, is its front of operations. There is an important consideration with reference to the direction of the front of operations and to changes it may receive, which I have dwelt upon in Article XX.

The fundamental principle requires, even when the forces are equal, that the front be less extensive than the enemy's,—especially if the front remains unchanged for some time. If your strategic positions are more closely connected than the enemy's, you can concentrate more rapidly and more easily than he can, and in this way the fundamental principle will be applied. If your positions are interior and central, the enemy cannot concentrate except by

---

1   The objective may be in some degree *political,*—especially in cases of wars of intervention in the affairs of another country; but it then really becomes geographical

passing by the mass of your divisions or by moving in a circle around them: he is then exactly in a condition not to be able to apply the fundamental principle, while it is your most obvious measure.

But if you are very weak and the enemy very strong, a central position, that may be surrounded on all sides by forces superior at every point, is untenable, unless the enemy's corps are very far separated from each other, as was the case with the allied armies in the Seven Years' War; or unless the central zone has a natural barrier on one or two of its sides, like the Rhine, the Danube, or the Alps, which would prevent the enemy from using his forces simultaneously. In case of great numerical inferiority it is, nevertheless, wiser to maneuver upon one of the extremities than upon the center of the enemy's line, especially if his masses are sufficiently near to be dangerous to you.

It was stated above that strategy, besides indicating the decisive points of a theater of war, requires two things:–1st, that the principal mass of the force be moved against fractions of the enemy's, to attack them in succession; 2d, that the best direction of movement be adopted,—that is to say, one leading straight to the decisive points already known, and afterward upon secondary points.

To illustrate these immutable principles of strategy, I will give a sketch of the operations of the French at the close of 1793. (See Plate III.)

It will be recollected that the allies had ten principal corps on the frontier of France from the Rhine to the North Sea.

The Duke of York was attacking Dunkirk. (No. 1.)

Marshal Freytag was covering the siege. (No. 2.)

The Prince of Orange was occupying an intermediate position at Menin. (No. 3.)

The Prince of Coburg, with the main army, was attacking Maubeuge, and was guarding the space between that place and the Scheldt by strong detachments. (No. 4.)

Clairfayt was covering the siege. (No. 5.)

Benjouski was covering Charleroi and the Meuse, toward Thuin and Charleroi, the fortifications of which were being rebuilt. (No. 6.)

Another corps was covering the Ardennes and Luxembourg. (No. 7.)

The Prussians were besieging Landau. (No. 8.)

The Duke of Brunswick was covering the siege in the Vosges. (No. 9.)

General Wurmser was observing Strasbourg and the army of the Rhine. (No. 10.)

The French, besides the detachments in front of each of the hostile corps, had five principal masses in the camps of Lille, Douai, Guise, Sarre Louis, and Strasbourg, (a, b, c, d, e.) A strong reserve, (g,) composed of the best troops drawn from the camps of the northern frontier, was intended to be thrown upon all the points of the enemy's line in succession, assisted by the troops already in the neighborhood, (i, k, l, m.)

This reserve; assisted by the divisions of the camp of Cassel near Dunkirk, commenced its operations by beating corps 1 and 2, under the Duke of York; then that of the Dutch, (No. 3,) at Menin; next that of Clairfayt, (5,) before Maubeuge; finally, joining the army of the Moselle toward Sarre Louis, it beat the Duke of Brunswick in the Vosges, and, with the assistance of the army of the Rhine, (f,) drove Wurmser from the lines of Wissembourg.

The general principle was certainly well applied, and every similar operation will be praiseworthy. But, as the Austrians composed half the allied forces, and they had their lines of retreat from the points 4, 5, and 6 upon the Rhine, it is evident that if the French had collected three of their large corps in order to move them against Benjouski at Thuin, (No. 6,) and then fallen upon the Prince of Coburg's left by the Charleroi road, they would have thrown the imperial army upon the North Sea, and would have obtained immense results.

The Committee of Public Safety deemed it a matter of great importance that Dunkirk should not be permitted to fell into the hands of the English. Besides this, York's corps, encamped on the downs, might be cut off and thrown upon the sea; and the disposable French masses for this object were at Douai, Lille, and Cassel: so that there were good reasons for commencing operations by attacking the English. The principal undertaking failed, because Houchard did not appreciate the strategic advantage he had, and did not know how to act on the line of retreat of the Anglo-Hanoverian army. He was guillotined, by way of punishment, although he saved Dunkirk; yet he failed to cut off the English as he might have done.

It will be observed that this movement of the French reserve along the whole front was the cause of five victories, neither of which had decisive results, *because the attacks were made in front*, and because, when the cities were relieved, the allied armies not being cut through, and the French reserve moving on to the different points in succession, none of the victories was pushed to its legitimate consequences. If the French had based themselves upon the five fortified towns on the Meuse, had collected one hundred thousand men by bold and rapid marches, had fallen upon the center of those separated corps, had crushed Benjouski, assailed the Prince of Coburg in his rear, beaten him, and pursued him vigorously as Napoleon pursued at Ratisbon, and as he wished to do at Ligny in 1815, the result would have been very different.

I have mentioned this example, as it illustrates very well the two important points to be attended to in the strategic management of masses of troops; that is, their employment at different points in succession and at decisive points.[1]

I   The operations mentioned show the advantage of employing masses at the decisive point, not because it was done in 1793, but because it was not done. If Napoleon had been in Carnot's place, he would have fallen with all his force upon Charleroi, whence be would have attacked the left of the Prince of Coburg and cut his line of retreat. Let any one compare the results of Carnot's half-skillful operations with the wise maneuvers of Saint-Bernard and Jena, and be convinced.

Every educated military man will be impressed by the truths educed, and will be convinced that the excellence of maneuvers will depend upon their conforming to the principle already insisted upon; that is to say, the great part of the force must be moved against one wing or the center, according to the position of the enemy's masses. It is of importance in battles to calculate distances with still greater accuracy; for the results of movements on the battle-field following them more rapidly than in the case of strategic maneuvers, every precaution must be taken to avoid exposing any part of the line to a dangerous attack from the enemy, especially if he is compactly drawn up. Add to these things calmness during the action; the ability to choose positions for fighting battles in the manner styled the defensive with *offensive returns,* (Art. XXX.;) the simultaneous employment of the forces in striking the decisive blow, the faculty of arousing the soldiers and moving them forward at opportune moments; and we have mentioned every thing which can assist, as far as the general is concerned, in assuring victories, and every thing which will constitute him a skillful tactician.

It is almost always easy to determine the decisive point of a field of battle, but not so with the decisive moment; and it is precisely here that genius and experience are every thing, and mere theory of little value.

It is important, also, to consider attentively Article XLII., which explains how a general may make a small number of suppositions as to what the enemy may or can do, and as to what course of conduct he shall himself pursue upon those hypotheses. He may thus accustom himself to be prepared for any eventuality.

I must also call attention to Article XXVIII., upon great detachments. These are necessary evils, and, if not managed with great care, may prove ruinous to the best armies. The essential rules on this point are, to make as few detachments as possible, *to have them readily movable,* to draw them back to the main body as soon as practicable, and to give them good instructions for avoiding disasters.

I have nothing to say relative to the first two chapters on military policy; for they are themselves nothing more than a brief summary of this part of the art of war, which chiefly concerns statesmen, but should be thoroughly understood by military men. I will, however, invite special attention to Article XIV., relating to the command of armies or to the choice of generals-in-chief,—a subject worthy the most anxious care upon the part of a wise government; for upon it often depends the safety of the nation.

We may be confident that a good strategist will make a good chief of staff for an army; but for the command in chief is required a man of tried qualities, of high character and known energy. The united action of two such men as commander-in-chief and chief of staff, when a great captain of the first order cannot be had, may produce the most brilliant results.

NOTE

UPON
THE MEANS OF ACQUIRING A GOOD STRATEGIC COUP-D'OEIL

*T*he study of the principles of strategy can produce no valuable practical results if we do nothing more than keep them in remembrance, never trying to apply them, with map in hand, to hypothetical wars, or to the brilliant operations of great captains. By such exercises may be procured a rapid and certain strategic *coup-d'oeil*,—the most valuable characteristic of a good general, without which he can never put in practice the finest theories in the world.

When a military man who is a student of his art has become fully impressed by the advantages procured by moving a strong mass against successive fractions of the enemy's force, and particularly when he recognizes the importance of constantly directing the main efforts upon decisive points of the theater of operations, he will naturally desire to be able to perceive at a glance what are these decisive points. I have already, in Chapter III., page 70, of the preceding Summary, indicated the simple means by which this knowledge may be obtained. There is, in fact, one truth of remarkable simplicity which obtains in all the combinations of a methodical war. It is this:—*in every position a general may occupy, he has only to decide whether to operate by the right, by the left, or by the front.*

To be convinced of the correctness of this assertion, let us first take this general in his private office at the opening of the war. His first care will be to choose that zone of operations which will give him the greatest number of chances of success and be the least dangerous for him in case of reverse. As no theater of operations can have more than three zones, (that of the right, that of the center, and that of the left,) and as I have in Articles from XVII. to XXII. pointed out the manner of perceiving the advantages and dangers of these zones, the choice of a zone of operations will be a matter of no difficulty.

When the general has finally chosen a zone within which to operate with the principal portion of his forces, and when these forces shall be established in that zone, the army will have a front of operations toward the hostile army, which will also have one. Now, these fronts of operations will each have its right, left, and center. It only remains, then, for the general to decide upon which of these directions he can injure the enemy most,—for this will always be the best, especially if he can move upon it without endangering his own communications. I have dwelt upon this point also in the preceding Summary.

Finally, when the two armies are in presence of each other upon the field of battle where the decisive collision is to ensue, and are upon the point of coming to blows, they will each have a right, left, and center; and it remains for the general to decide still between these three directions of striking.

Let us take, as an illustration of the truths I have mentioned, the theater of operations, already referred to, between the Rhine and the North Sea. (See Fig. 39.)

Although this theater presents, in one point of view, four geographical sections,—viz.: the space between the Rhine and the Moselle, that between the Moselle and the Meuse, that between the Meuse and the Scheldt, and that between the last river and the sea,—it is nevertheless true that an army of which A A is the base and B B the front of operations will have only three general directions to choose from; for the two spaces in the center will form a single central zone, as it will always have one on the right and another on the left.

Figure 39

255

The army B B, wishing to take the offensive against the army CC, whose base was the Rhine, would have three directions in which to operate. If it maneuvered by the extreme right, descending the Moselle, (toward D,) it would evidently threaten the enemy's line of retreat toward the Rhine; but he, concentrating the mass of his forces toward Luxembourg, might fall upon the left of the army D and compel it to change front and fight a battle with its rear toward the Rhine, causing its ruin if seriously defeated.

If, on the contrary, the army B wished to make its greatest effort upon the left, (toward E,) in order to take advantage of the finely-fortified towns of Lille and Valenciennes, it would be exposed to inconveniences still more serious than before. For the army CC, concentrating in force toward Audenarde, might fall on the right of B, and, outflanking this wing in the battle, might throw it upon the impassable country toward Antwerp between the Scheldt and the sea,—where there would remain but two things for it to do: either to surrender at discretion, or cut its way through the enemy at the sacrifice of half its numbers.

It appears evident, therefore, that the left zone would be the most disadvantageous for army B, and the right zone would be inconvenient, although somewhat favorable in a certain point of view. The central zone remains to be examined. This is found to possess all desirable advantages, because the army B might move the mass of its force toward Charleroi with a view of cutting through the immense front of operations of the enemy, might overwhelm his center, and drive the right back upon Antwerp and the Lower Scheldt, without seriously exposing its own communications.

When the forces are chiefly concentrated upon the most favorable zone, they should, of course, have that direction of movement toward the enemy's front of operations which is in harmony with the chief object in view. For example, if you shall have operated by your right against the enemy's left, with the intention of cutting off the greater portion of his army from its base of the Rhine, you should certainly continue to operate in the same direction; for if you should make your greatest effort against the right of the enemy's front, while your plan was to gain an advantage over his left, your operations could not result as you anticipated, no matter how well they might be executed. If, on the contrary, you had decided to take the left zone, with the intention of crowding the enemy back upon the sea, you ought constantly to maneuver by your right in order to accomplish your object; for if you maneuvered by the left, yourself and not the enemy would be the party thrown back upon the sea in case of a reverse.

Applying these ideas to the theaters of the campaigns of Marengo, Ulm, and Jena, we find the same three zones, with this difference, that in those campaigns the central direction was not the best. In 1800, the direction of the left led straight to the left bank of the Po, on the line of retreat of Melas; in 1805, the left zone was the one which led by the way of Donauwerth to the extreme right, and the line of retreat of Mack; in 1806, however, Napoleon could reach the Prussian line of retreat by the right zone, filing off from Bamberg toward Gera.

In 1800, Napoleon had to choose between a line of operations on the right, leading to the sea-shore toward Nice and Savona, that of the center, leading by Mont-Cenis toward Turin, and that of the left, leading to the line of communications of Melas, by way of Saint-Bernard or the Simplon. The first two directions had nothing in their favor, and the right might have been very dangerous,—as, in fact, it proved to Massena, who was forced back to Genoa and there besieged. The decisive direction was evidently that by the left.

I have said enough to explain my ideas on this point.

The subject of battles is somewhat more complicated; for in the arrangements for these there are both strategical and tactical considerations to be taken into account and harmonized. A position for battle, being necessarily connected with the line of retreat and the base of operations, must have a well-defined strategic direction; but this direction must also depend somewhat upon the character of the ground and the stations of the troops of both parties to the engagement: these are tactical considerations. Although an army usually takes such a position for a battle as will keep its line of retreat behind it, sometimes it is obliged to assume a position parallel to this line. In such a case it is evident that if you fall with overwhelming force upon the wing nearest the line of retreat, the enemy may be cut off or destroyed, or, at least, have no other chance of escape than in forcing his way through your line.

I will here mention as illustrations the celebrated battle of Leuthen in 1757, of which I have given an account in the history of Frederick's wars, and the famous days of Krasnoi, in the retreat from Moscow in 1812.

Figure 40

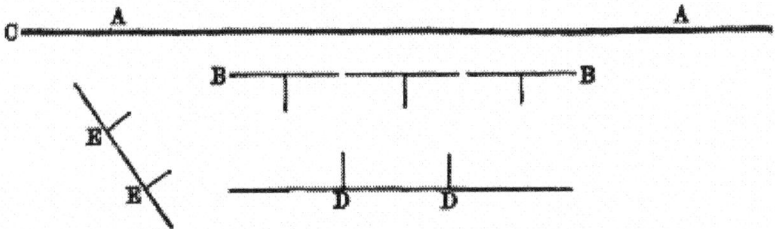

The annexed figure (40) explains the combination at Krasnoi. The line A A is Napoleon's line of retreat toward C. He took the position B B to cover his line. It is evident that the principal mass of Koutousoff's army D D should have moved to E E in order to fall on the right of the French, whose army would have been certainly destroyed if it had been anticipated at C; for everybody knows in what a state it was while thus fifteen hundred miles from its true base.

There was the same combination at Jemmapes, where Dumouriez, by outflanking the Austrian left, instead of attacking their right, would have entirely cut them off from the Rhine.

At the battle of Leuthen Frederick overwhelmed the Austrian left, which was in the direction of their line of retreat; and for this reason the right wing was obliged to take refuge in Breslau, where it capitulated a few days later.

In such cases there is no cause for hesitation. The decisive point is that wing of the enemy which is nearest his line of retreat, and this line you must seize while protecting your own.

When an enemy has one or two lines of retreat perpendicular to and behind his position of battle, it will generally be best to attack the center, or that wing where the obstacles of the ground shall be the least favorable for the defense; for in such a case the first consideration is to gain the battle, without having in view the total destruction of the enemy. That depends upon the relative numerical strength, the *morale* of the two armies, and other circumstances, with reference to which no fixed rules can be laid down.

Finally, it happens sometimes that an army succeeds in seizing the enemy's line of retreat before fighting a battle, as Napoleon did at Marengo, Ulm, and Jena. The decisive point having in such case been secured by skillful marches before fighting, it only remains to prevent the enemy from forcing his way through your line. You can do nothing better than fight a parallel battle, as there is no reason for maneuvering against one wing more than the other. But for the enemy who is thus cut off the case is very different. He should certainly strike most heavily in the direction of that wing where he can hope most speedily to regain his proper line of retreat; and if he throws the mass of his forces there, he may save at least a large portion of them. All that he has to do is to determine whether this decisive effort shall be toward the right or the left.

It is proper for me to remark that the passage of a great river in the presence of a hostile army is sometimes an exceptional case to which the general rules will not apply. In these operations, which are of an exceedingly delicate character, the essential thing is to keep the bridges safe. If, after effecting the passage, a general should throw the mass of his forces toward the right or the left with a view of taking possession of some decisive point, or of driving his enemy back upon the river, whilst the latter was collecting all his forces in another direction to seize the bridges, the former army might be in a very critical condition in case of a reverse befalling it. The battle of Wagram is an excellent example in point,—as good, indeed, as could be desired. I have treated this subject in Article XXXVII., (pages 224 and following.)

A military man who clearly perceives the importance of the truths that have been stated will succeed in acquiring a rapid and accurate *coup-d'oeil*. It will be admitted, moreover, that a general who estimates them at their true value, and accustoms himself to their use, either in reading military history, or in hypothetical cases on maps, will seldom be in doubt, in real campaigns, what he ought to do; and even when his enemy attempts sudden and unexpected movements, he will always be ready with suitable measures for counteracting them, by constantly bearing in mind the few simple fundamental principles which should regulate all the operations of war.

Heaven forbid that I should pretend to lessen the dignity of the sublime art of war by reducing it to such simple elements! I appreciate thoroughly the difference between the directing principles of combinations arranged in the quiet of the closet, and that special talent which is indispensable to the individual who has, amidst the noise and confusion of battle, to keep a hundred thousand men co-operating toward the attainment of one single object. I know well what should be the character and talents of the general who has to make such masses move as one man, to engage them at the proper point simultaneously and at the proper moment, to keep them supplied with arms, provisions, clothing, and munitions. Still, although this special talent, to which I have referred, is indispensable, it must be granted that the ability to give wise direction to masses upon the best strategic points of a theater of operations is the most sublime characteristic of a great captain. How many brave armies, under the command of leaders who were also brave and possessed executive ability, have lost not only battles, but even empires, because they were moved imprudently in one direction when they should have gone in the other! Numerous examples might be mentioned; but I will refer only to Ligny, Waterloo, Bautzen, Dennewitz, Leuthen.

I will say no more; for I could only repeat what has already been said. To relieve myself in advance of the blame which will be ascribed to me for attaching too much importance to the application of the few maxims laid down in my writings, I will repeat what I was the first to announce:–" *that war is not an exact science, but a drama full of passion*; that the moral qualities, the talents, the executive foresight and ability, the greatness of character, of the leaders, and the impulses, sympathies, and passions of the masses, have a great influence upon it." I may be permitted also, after having written the detailed history of thirty campaigns and assisted in person in twelve of the most celebrated of them, to declare that I have not found a single case where these principles, correctly applied, did not lead to success.

As to the special executive ability and the well-balanced penetrating mind which distinguish the practical man from the one who knows only what others teach him, I confess that no book can introduce those things into a head where the germ does not previously exist by nature. I have seen many generals— marshals, even—attain a certain degree of reputation by talking largely of principles which they conceived incorrectly in theory and could not apply at all. I have seen these men intrusted with the supreme command of armies, and make the most extravagant plans, because they were totally deficient in good judgment and were filled with inordinate self-conceit. My works are not intended for such misguided persons as these, but my desire has been to facilitate the study of the art of war for careful, inquiring minds, by pointing out directing principles. Taking this view, I claim credit for having rendered valuable service to those officers who are really desirous of gaining distinction in the profession of arms.

Finally, I will conclude this short summary with one last truth:—

"The first of all the requisites for a man's success as a leader is, that he be perfectly brave. When a general is animated by a truly martial spirit and can communicate it to his soldiers, he may commit faults, but he will gain victories and secure deserved laurels."

SECOND APPENDIX
TO THE
SUMMARY OF THE ART OF WAR

ON THE FORMATION OF TROOPS FOR BATTLE

*H*appening to be in Paris, near the end of 1851, a distinguished person did me the honor to ask my opinion as to whether recent improvements in fire-arms would cause any great modifications in the manner of making war.

I replied that they would probably have an influence upon the details of tactics, but that, in great strategic operations and the grand combinations of battles, victory would, now as ever, result from the application of the principles which had led to the success of great generals in all ages,--of Alexander and Caesar as well as of Frederick and Napoleon. My illustrious interlocutor seemed to be completely of my opinion.

The heroic events which have recently occurred near Sebastopol have not produced the slightest change in my opinion. This gigantic contest between two vast intrenched camps, occupied by entire armies and mounting two thousand guns of the largest caliber, is an event without precedent, which will have no equal in the future; for the circumstances which produced it cannot occur again.

Moreover, this contest of cannon with ramparts, bearing no resemblance to regular pitched battles fought in the center of a continent, cannot influence in any respect the great combinations of war, nor even the tactics of battles.

The bloody battles of the Alma and Inkermann, by giving evidence of the murderous effect of the new fire-arms, naturally led me to investigate the changes which it might be necessary to make on this account in the tactics for infantry.

I shall endeavor to fulfill this task in a few words, in order to complete what was published on this point twenty years ago in the Summary of the Art of War.

The important question of the influence of musketry-fire in battles is not new: it dates from the reign of Frederick the Great, and particularly from the battle of Mollwitz, which he gained (it was said) because his infantry-soldiers, by the use of cylindrical rammers in loading their muskets, were able to fire three shots per minute more than their enemies.[I] The discussion which arose at this epoch between the partisans of the shallow and deep orders of formation for troops is known to all military students.

The system of deployed lines in three ranks was adopted for the infantry; the cavalry, formed in two ranks, and in the order of battle, was deployed upon the wings, or a part was held in reserve.

The celebrated regulation for maneuvers of 1791 fixed the deployed as the only order for battle: it seemed to admit the use of battalion-columns doubled on the center only in partial combats,--such as an attack upon an isolated post, a village, a forest, or small intrenchments.[II]

The insufficient instruction in maneuvers of the troops of the Republic forced the generals, who were poor tacticians, to employ in battle the system of columns supported by numerous skirmishers. Besides this, the nature of the countries which formed the theaters of operations--the Vosges, Alps, Pyrenees, and the difficult country of La Vendee--rendered this the only appropriate system. How would it have been possible to attack the camps of Saorgio, Figueras, and Mont-Cenis with deployed regiments?

In Napoleon's time, the French generally used the system of columns, as they were nearly always the assailants.

In 1807, I published, at Glogau in Silesia, a small pamphlet with the title of "Summary of the General Principles of the Art of War," in which I proposed to admit for the attack the system of lines formed of columns of battalions by divisions of two companies; in other words, to march to the attack in lines of battalions closed in mass or at half-distance, preceded by numerous skirmishers, and the columns being separated by intervals that may vary between that necessary for the deployment of a battalion and the minimum of the front of one column.

What I had recently seen in the campaigns of Ulm, Austerlitz, Jena, and Eylau had convinced me of the difficulty, if not the impossibility, of marching an army in deployed lines in either two or three ranks, to attack an enemy in position. It was this conviction which led me to publish the pamphlet above

---

I It is probable that Baron Jomini here refers to iron, instead of cylindrical, ramrods. Before 1730, all European troops used wooden ramrods; and the credit of the invention of iron ones is attributed by some to the Prince of Anhalt, and by others to Prince Leopold of Dessau. The Prussians were the first to adopt the iron ramrod, and at the date of the battle of Mollwitz (1741) it had not been introduced into the Austrian service.
Frederick did not adopt the cylindrical ramrod till 1777, thirty-six years after the battle of Mollwitz. The advantage of the cylindrical ramrod consisted in this,--that the soldier in loading saved the time necessary to turn the ramrod; but obviously this small economy of time could never have enabled him to load three times while the enemy loaded once,--all other things being equal.--Translators.
II Columns by battalions closed in mass seemed only to be intended to use in long columns on the march, to keep them closed, in order to facilitate their deployment.

referred to. This work attracted some attention, not only on account of the treatise on strategy, but also on account of what was said on tactics.

The successes gained by Wellington in Spain and at Waterloo with troops deployed in lines of two ranks were generally attributed to the murderous effect of the infantry-fire, and created doubt in some minds as to the propriety of the use of small columns; but it was not till after 1815 that the controversies on the best formation for battle wore renewed by the appearance of a pamphlet by the Marquis of Chambray.

In these discussions, I remarked the fatal tendency of the clearest minds to reduce every system of war to absolute forms, and to cast in the same mold all the tactical combinations a general may arrange, without taking into consideration localities, moral circumstances, national characteristics, or the abilities of the commanders. I had proposed to use lines of small columns, especially in the attack: I never intended to make it an exclusive system, particularly for the defense.

I had two opportunities of being convinced that this formation was approved of by the greatest generals of our times. The first was at the Congress of Vienna, in the latter part of 1814: the Archduke Charles observed "that he was under great obligations for the summary I had published in 1807, which General Walmoden had brought to him in 1808 from Silesia." At the beginning of the war of 1809, the prince had not thought it possible to apply the formation which I had proposed; but at the battle of Essling the contracted space of the field induced him to form a part of his army in columns by battalions, (the landwehr particularly,) and they resisted admirably the furious charges of the cuirassiers of General d'Espagne, which, in the opinion of the archduke, they could not have done if they had been deployed.

At the battle of Wagram, the greater part of the Austrian line was formed in the same way as at Essling, and after two days of terrible fighting the archduke abandoned the field of battle, not because his army was badly beaten, but because his left was outflanked and thrown back so as to endanger his line of retreat on Hungary. The prince was satisfied that the firm bearing of his troops was in part due to this mixture of small columns with deployed battalions.

The second witness is Wellington; although his evidence is, apparently, not so conclusive. Having been presented to him at the Congress of Verona in 1823, I had occasion to speak to him on the subject of the controversies to which his system of formation for battle (a system to which a great part of his success had been attributed) had given rise. He remarked that he was convinced the manner of the attack of the French upon him, in columns more or less deep, was very dangerous against a solid, well-armed infantry having confidence in its fire and well supported by artillery and cavalry. I observed to the duke that these deep columns were very different from the small columns which I proposed,--a formation which insures in the attack steadiness, force, and mobility, while deep masses afford no greater mobility and force than a deployed line, and are very much more exposed to the ravages of artillery.

I asked the illustrious general if at Waterloo he had not formed the Hanoverian, Brunswick, and Belgian troops in columns by battalions. He answered, "Yes; because I could not depend upon them so well as upon the English." I replied that this admission proved that he thought a line formed of columns by battalions was more firm than long deployed lines. He replied, "They are certainly good, also; but their use always depends upon the localities and the spirit of the troops. A general cannot act in the same manner under all circumstances."

To this illustrious evidence I might add that Napoleon himself, in the campaign of 1813, prescribed for the attack the formation of the infantry in columns by divisions of two companies in two ranks, as the most suitable,-- which was identically what I had proposed in 1807.

The Duke of Wellington also admitted that the French columns at Waterloo, particularly those of their right wing, were not small columns of battalions, but enormous masses, much more unwieldy and much deeper.

If we can believe the Prussian accounts and plans of the battle, it would seem that Ney's four divisions were formed in but four columns, at least in their march to the attack of La Haye Sainte and the line extending from this farm to the Papelotte. I was not present; but several officers have assured me that at one time the troops were formed in columns by divisions of two brigades each, the battalions being deployed behind each other at six paces' interval.

This circumstance demonstrates how much is wanting in the military terms of the French. We give the same name of *division* to masses of four regiments and to fractions of a battalion of two companies each, —which is absurd. Let us suppose, for example, that Napoleon had directed on the 18th of June, 1815, the formation of the line in columns by divisions and by battalions, intending that the regulation of 1813 should be followed. His lieutenants might naturally have understood it very differently, and, according to their interpretation of the order, would have executed one of the following formations:–

1. Either the four divisions of the right wing would have been formed in four large masses, each one of eight or twelve battalions, (according to the strength of the regiments,) as is indicated in this figure for eight battalions.[1]

---

[1]   We suppose each regiment to consist of two battalions: if there should be three in each regiment, the deep column would then consist of twelve lines of either twenty-four or thirty-six ranks, while in the next figure there would be twelve battalions on the line instead of eight, the depth not being increased.

264

2. Or each division would have been formed in eight or twelve columns of battalions by divisions of two platoons or companies, according to the system I have proposed, as in this figure, viz.:–

I do not mean to assert positively that this confusion of words led to the deep masses at Waterloo; but it might have done so; and it is important that in every language there should be two different terms to express two such different things as a *division* of twelve battalions and a *division* of a quarter of a battalion.

Struck with what precedes, I thought it proper to modify my Summary already referred to, which was too concise, and in my revision of it I devoted a chapter to the discussion of the advantages and disadvantages of the different formations for battle. I also added some considerations relative to a mixed system used at Eylau by General Benningsen, which consisted in forming a regiment of three battalions by deploying the central one, the other two being in column on the wings.

After these discussions, I drew the conclusions:–

1. That Wellington's system was certainly good for the defensive.

2. That the system of Benningsen might, according to circumstances, be as good for the offensive as for the defensive, since it was successfully used by Napoleon at the passage of the Tagliamento.

3. That the most skillful tactician would experience great difficulty in marching forty or fifty deployed battalions in two or three ranks over an interval of twelve or fifteen hundred yards, preserving sufficient order to attack an enemy in position with any chance of success, the front all the while being played upon by artillery and musketry.

I have never seen any thing of the kind in my experience. I regard it as impossible, and am convinced that such a line could not advance to the attack in sufficiently good order to have the force necessary for success.

Napoleon was in the habit of addressing his marshals in these terms:–"Take your troops up in good order, and make a vigorous assault upon the enemy." I ask, what means is there of carrying up to the assault of an enemy forty or fifty deployed battalions as a whole in good order? They will reach the enemy in detachments disconnected from each other, and the commander cannot exercise any control over the mass as a whole.

I saw nothing of this kind either at Ulm, Jena, Eylau, Bautzen, Dresden, Culm, or Leipsic; neither did it occur at Austerlitz, Friedland, Katzbach, or Dennewitz.

I am not aware that Wellington, in any of his battles, ever marched in deployed lines to the attack of an enemy in position. He generally awaited the attack. At Vittoria and Toulouse he gained the victory by maneuvers against the flanks; and at Toulouse Soult's right wing was beaten while descending the heights to attack. Even at Waterloo, what fate would have befallen the English army if, leaving the plateau of Mont Saint-Jean, it had marched in deployed order to attack Napoleon in position on the heights of La Belle Alliance?

I will be pardoned for these recapitulations, as they seem to be necessary to the solution of a question which has arisen since my Summary of the Art of War was written.

Some German generals, recognizing fully the advantages derived in 1813 from the system of columns of battalions, have endeavored to add to its value by dividing up the columns and increasing their number, so as to make them more shallow and to facilitate their deployment. With this view, they propose, instead of forming four divisions or companies one behind the other, to place them beside each other, not deployed, but in small columns. That is, if the battalion consists of four companies of two hundred and forty men each, each company is to be divided into four sections of sixty each: one of these sections will be dispersed as skirmishers, and the other three, in two ranks, will form a small column; so that the battalion, instead of forming one column, will form four, and the regiment of three battalions will form twelve small columns instead of three--

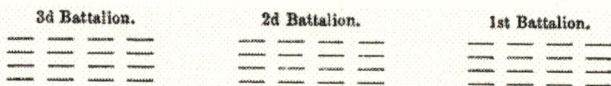

3d Battalion.	2d Battalion.	1st Battalion.

It is certain that it would be easier to march such a line against the enemy than if deployed; but these diminutive columns of sixty skirmishers and one hundred and eighty men in the ranks would never present the same order and solidity as a single column of a battalion. Still as the system has some advantages, it deserves a trial; and, indeed, it has already been practiced in Prussia and Austria.

The same formation applies equally to battalions of six or eight companies. In this case the battalion would not be formed by companies, but by divisions of two companies,--that is, in three or four columns, according to the number of companies.

Two serious inconveniences appear to me to attach to each of these formations. If vigorously charged by cavalry, these small subdivisions would be in great danger; and even in attacking the enemy's line, if driven back and pursued, disorder would be more likely to occur than in the columns of battalions. Still, either of them may be employed, according to circumstances, localities, and the *morale* of the troops. Experience alone can assign to each its proper value. I am not aware whether the Austrians applied these columns of companies at Custozza and Novara, or whether these maneuvers have only been practiced in their camps of instruction.

Be that as it may, there is another not less important question to be considered:—

"Will the adoption of the rifled small-arms and improved balls bring about any important changes in the formation for battle and the now recognized principles of tactics?"

If these arms aided the allies at the Alma and Inkermann, it was because the Russians were not provided with them; and it must not be forgotten that in a year or two all armies will alike be furnished with them, so that in future the advantage will not be confined to one side.

What change will it make in tactics?

Will whole armies be deployed as skirmishers, or will it not still be necessary to preserve either the formation of lines deployed in two or three ranks, or lines of battalions in columns?

Will battles become mere duels with the rifle, where the parties will fire upon each other, without maneuvering, until one or the other shall retreat or be destroyed?

What military man will reply in the affirmative?

It follows, therefore, that, to decide battles, maneuvers are necessary, and victory will fall to the general who maneuvers most skillfully; and he cannot maneuver except with deployed lines or lines of columns of battalions, either whole or subdivided into columns of one or two companies. To attempt to prescribe by regulation under what circumstances either of these systems is to be applied would be absurd.

If a general and an army can be found such that he can march upon the enemy in a deployed line of forty or fifty battalions, then let the shallow order be adopted, and the formation in columns be confined to the attack of isolated posts; but I freely confess that I would never accept the command of an army under this condition. The only point for a regulation for the formation for battle is to forbid the use of very deep columns, because they are heavy, and difficult to move and to keep in order. Besides, they are so much exposed to artillery that their destruction seems inevitable, and their great depth does not increase in any respect their chances of success.

If the organization of an army were left to me, I would adopt for infantry the formation in two ranks, and a regimental organization according with the formation for battle. I would then make each regiment of infantry to consist of three battalions and a depot. Each battalion should consist of six companies, so that when in column by division the depth would be three divisions or six ranks.

This formation seems most reasonable, whether it is desired to form the battalion in columns of attack by divisions on the center of each battalion, or on any other division.

The columns of attack, since the depth is only six ranks, would not be so much exposed to the fire of artillery, but would still have the mobility necessary to take the troops up in good order and launch them upon the enemy with great force. The deployment of these small columns could be executed with great ease and promptitude; and for the forming of a square a column of three divisions in depth would be preferable in several respects to one of four or six divisions.

In the Russian service each battalion consists of four companies of two hundred and fifty men each; each company being as strong as a division in the French organization. The maneuver of double column on the center is not practicable, since the center is here merely an interval separating the second and third companies. Hence the column must be simple, not on the center, but on one of the four companies. Something analogous to the double column on the center would be attained by forming the first and fourth companies behind the second and third respectively; but then the formation would be in two lines rather than in column; and this is the reason why I would prefer the organization of the battalion in six companies or three divisions.

By dividing each of the four companies into two platoons, making eight in all, the formation of *double column on the center* might be made on the fourth and fifth platoons as the leading division; but then each division would be composed of two platoons belonging to different companies, so that each captain would have half of the men of his company under the command of another officer, and half of his own division would be made up of another company.

Such an arrangement in the attack would be very inconvenient; for, as the captain is the real commander, father, and judge of the men of his own company, he can always obtain more from them in the way of duty than any stranger. In addition, if the double column should meet with a decided repulse, and it should be necessary to reform it in line, it would be difficult to prevent disorder, the platoons being obliged to run from one side to the other to find their companies. In the French system, where each battalion consists of eight companies, forming as many platoons at drill, this objection does not exist, since each company is conducted by its own captain. It is true that there will be two captains of companies in each division; but this will be rather an advantage than the reverse, since there will be a rivalry and emulation between the two captains and their men, which will lead to greater display of bravery: besides, if necessary, the senior captain is there, to command the division as a whole.

It is time to leave these secondary details and return to the important question at issue.

Since I have alluded to the system adopted by Wellington, it is proper to explain it so that it can be estimated at its true value in the light of historical events.

In Spain and Portugal, particularly, Wellington had under his command a mass of troops of the country, in which he placed but little confidence in regular formation in a pitched battle, on account of their want of instruction and discipline, but which were animated by a lively hatred of the French and formed bodies of skirmishers useful in harassing the enemy. Having learned by experience the effects of the fury and impetuosity of the French columns when led by such men as Massena and Ney, Wellington decided upon wise means of weakening this impetuosity and afterward securing a triumph over it. He chose positions difficult to approach, and covered all their avenues by swarms of Spanish and Portuguese riflemen, who were skilled in taking advantage of the inequalities of the ground; he placed a part of his artillery on the tactical crest of his position, and a part more to the rear, and riddled the advancing columns with a murderous artillery and musketry fire, while his excellent English infantry, sheltered from the fire, were posted a hundred paces in rear of the crest, to await the arrival of these columns; and when the latter appeared on the summit, wearied, out of breath, decimated in numbers, they were received with a general discharge of artillery and musketry and immediately charged by the infantry with the bayonet.

This system, which was perfectly rational and particularly applicable to Spain and Portugal, since he had there great numbers of this kind of troops and there was a great deal of rough ground upon which they could be useful as marksmen, needed some modifications to make it applicable to Belgium. At Waterloo the duke took his position on a plateau with a gentle slope like a glacis, where his artillery had a magnificent field of fire, and where it produced a terrible effect: both flanks of this plateau were well protected. Wellington, from the crest of the plateau, could discover the slightest movement in the French army, while his own were hidden; but, nevertheless, his system would not have prevented his losing the battle if a number of other circumstances had not come to his aid.

Every one knows more or less correctly the events of this terrible battle, which I have elsewhere impartially described. I demonstrated that its result was due neither to the musketry-fire nor to the use of deployed lines by the English, but to the following accidental causes, viz.:–

1. To the mud, which rendered the progress of the French in the attack painful and slow, and caused their first attacks to be less effective, and prevented their being properly sustained by the artillery.

2. To the original formation of very deep columns on the part of the French, principally on the right wing.

3. To the want of unity in the employment of the three arms: the infantry and cavalry made a number of charges alternating with each other, but they were in no case simultaneous.

4. Finally and chiefly, to the unexpected arrival of the whole Prussian army at the decisive moment on the right flank, if not the rear, of the French.

Every experienced military man will agree that, in spite of the mud and the firmness of the English infantry, if the mass of the French infantry had been thrown on the English in columns of battalions immediately after the great charge of cavalry, the combined army would have been broken and forced back on Antwerp. Independently of this, if the Prussians had not arrived, the English would have been compelled to retreat; and I maintain that this battle cannot justly be cited as proof of the superiority of musketry-fire over well-directed attacks in columns.

From all these discussions we may draw the following conclusions, viz.:–

1. That the improvements in fire-arms will not introduce any important change in the manner of taking troops into battle, but that it would be useful to introduce into the tactics of infantry the formation of columns by companies, and to have a numerous body of good riflemen or skirmishers, and to exercise the troops considerably in firing. Those armies which have whole regiments of light infantry may distribute them through the different brigades; but it would be preferable to detail sharp-shooters alternately in each company as they are needed, which would be practicable when the troops are accustomed to firing: by this plan the light-infantry regiments could be employed in the line with the others; and should the number of sharp-shooters taken from the companies be at any time insufficient, they could be reinforced by a battalion of light infantry to each division.

2. That if Wellington's system of deployed lines and musketry-fire be excellent for the defense, it would be difficult ever to employ it in an attack upon an enemy in position.

3. That, in spite of the improvements of fire-arms, two armies in a battle will not pass the day in firing at each other from a distance: it will always be necessary for one of them to advance to the attack of the other.

4. That, as this advance is necessary, success will depend, as formerly, upon the most skillful maneuvering according to the principles of grand tactics, which consist in this, viz.: in knowing how to direct the great

mass of the troops at the proper moment upon the decisive point of the battle-field, and in employing for this purpose the simultaneous action of the three arms.

5. That it would be difficult to add much to what has been said on this subject in Chapters IV. and V.; and that it would be unreasonable to define by regulation an absolute system of formation for battle.

6. That victory may with much certainty be expected by the party taking the offensive when the general in command possesses the talent of taking his troops into action in good order and of boldly attacking the enemy, adopting the system of formation best adapted to the ground, to the spirit and quality of his troops, and to his own character.

Finally, I will terminate this article with the following remark: That war, far from being an exact science, is a terrible and impassioned drama, regulated, it is true, by three or four general principles, but also dependent for its results upon a number of moral and physical complications.

SKETCH OF THE PRINCIPAL MARITIME EXPEDITIONS

I have thought it proper to give here an account of the principal maritime expeditions, to be taken in connection with maxims on descents.

The naval forces of Egypt, Phoenicia, and Rhodes are the earliest mentioned in history, and of them the account is confused. The Persians conquered these nations, as well as Asia Minor, and became the most formidable power on both land and sea.

About the same time the Carthaginians, who were masters of the coast of Mauritania, being invited by the inhabitants of Cadiz, passed the straits, colonized Boetica and took possession of the Balearic Isles and Sardinia, and finally made a descent on Sicily.

The Greeks contended against the Persians with a success that could not have been expected,--although no country was ever more favorably situated for a naval power than Greece, with her fifty islands and her great extent of coast.

The merchant marine of Athens produced her prosperity, and gave her the naval power to which Greece was indebted for her independence. Her fleets, united with those of the islands, were, under Themistocles, the terror of the Persians and the rulers of the East. They never made grand descents, because their land-forces were not in proportion to their naval strength. Had Greece been a united government instead of a confederation of republics, and had the navies of Athens, Syracuse, Corinth, and Sparta been combined instead of fighting among each other, it is probable that the Greeks would have conquered the world before the Romans.

If we can believe the exaggerated traditions of the old Greek historians, the famous army of Xerxes had not less than four thousand vessels; and this number is astonishing, even when we read the account of them by Herodotus. It is more difficult to believe that at the same time, and by a concerted movement, five thousand other vessels landed three hundred thousand Carthaginians in Sicily, where they were totally defeated by Gelon on the same day that Themistocles destroyed the fleet of Xerxes at Salamis. Three other expeditions, under Hannibal, Imilcon, and Hamilcar, carried into Sicily from one hundred to one hundred and fifty thousand men: Agrigentum and Palermo were taken, Lilybaeum was founded, and Syracuse besieged twice. The third time Androcles, with fifteen thousand men, landed in Africa, and made Carthage tremble. This contest lasted one year and a half.

Alexander the Great crossed the Hellespont with only fifty thousand men: his naval force was only one hundred and sixty sail, while the Persians had four hundred; and to save his fleet Alexander sent it back to Greece.

After Alexander's death, his generals, who quarreled about the division of the empire, made no important naval expedition.

Pyrrhus, invited by the inhabitants of Tarentum and aided by their fleet, landed in Italy with twenty-six thousand infantry, three thousand horses, and the first elephants which had been seen in Italy. This was two hundred and eighty years before the Christian era.

Conqueror of the Romans at Heraclea and Ascoli, it is difficult to understand why he should have gone to Sicily at the solicitation of the Syracusans to expel the Carthaginians. Recalled, after some success, by the Tarentines, he recrossed the straits, harassed by the Carthaginian fleet: then, reinforced by the Samnites or Calabrians, he, a little too late, concluded to march on Rome. He in turn was beaten and repulsed on Beneventum, when he returned to Epirus with nine thousand men, which was all that remained of his force.

Carthage, which had been prospering for a long time, profited by the ruin of Tyre and the Persian empire.

The Punic wars between Carthage and Rome, now the preponderating power in Italy, were the most celebrated in the maritime annals of antiquity. The Romans were particularly remarkable for the rapidity with which they improved and increased their marine. In the year 264 B.C. their boats or vessels were scarcely fit to cross to Sicily; and eight years after found Regulus conqueror at Ecnomos, with three hundred and forty large vessels, each with three hundred rowers and one hundred and twenty combatants, making in all one hundred and forty thousand men. The Carthaginians, it is said, were stronger by twelve to fifteen thousand men and fifty vessels.

The victory of Ecnomos--perhaps more extraordinary than that of Actium--was the first important step of the Romans toward universal empire. The subsequent descent in Africa consisted of forty thousand men; but the greater part of this force being recalled to Sicily, the remainder was overthrown, and Regulus, being made prisoner, became as celebrated by his death as by his famous victory.

The great fleet which was to avenge him was successful at Clypea, but was destroyed on its return by a storm; and its successor met the same fate at Cape

Palinuro. In the year 249 B.C. the Romans were defeated at Drepanum, and lost twenty-eight thousand men and more than one hundred vessels. Another fleet, on its way to besiege Lilybaeum, in the same year, was lost off Cape Pactyrus.

Discouraged by this succession of disasters, the Senate at first resolved to renounce the sea; but, observing that the power of Sicily and Spain resulted from their maritime superiority, it concluded to arm its fleets again, and in the year 242 Lutatius Catullus set out with three hundred galleys and seven hundred transports for Drepanum, and gained the battle in the AEgates Islands, in which the Carthaginians lost one hundred and twenty vessels. This victory brought to a close the first Punic war.

The second, distinguished by Hannibal's expedition to Italy, was less maritime in its character. Scipio, however, bore the Roman eagles to Cartagena, and by its capture destroyed forever the empire of the Carthaginians in Spain. Finally, he carried the war into Africa with a force inferior to that of Regulus; but still he succeeded in gaining the battle of Zama, imposing a shameful peace on Carthage and burning five hundred of her ships. Subsequently Scipio's brother crossed the Hellespont with twenty-five thousand men, and at Magnesia gained the celebrated victory which surrendered to the mercy of the Romans the kingdom of Antiochus and all Asia. This expedition was aided by a victory gained at Myonnesus in Ionia, by the combined fleets of Rome and Rhodes, over the navy of Antiochus.

From this time Rome had no rival, and she continued to add to her power by using every means to insure to her the empire of the sea. Paulus Emilius in the year 168 B.C. landed at Samothrace at the head of twenty-five thousand men, conquered Perseus, and brought Macedonia to submission.

Twenty years later, the third Punic war decided the fate of Carthage. The important port of Utica having been given up to the Romans, an immense fleet was employed in transporting to this point eighty thousand foot-soldiers and four thousand horses; Carthage was besieged, and the son of Paulus Emilius and adopted son of the great Scipio had the glory of completing the victory which Emilius and Scipio had begun, by destroying the bitter rival of his country.

After this triumph, the power of Rome in Africa, as well as in Europe, was supreme; but her empire in Asia was for a moment shaken by Mithridates. This powerful king, after seizing in succession the small adjacent states, was in command of not less than two hundred and fifty thousand men, and of a fleet of four hundred vessels, of which three hundred were decked. He defeated the three Roman generals who commanded in Cappadocia, invaded Asia Minor and massacred there at least eighty thousand Roman subjects, and even sent a large army into Greece.

Sylla landed in Greece with a reinforcement of twenty-five thousand Romans, and retook Athens; but Mithridates sent in succession two large armies by the Bosporus and the Dardanelles: the first, one hundred thousand strong, was destroyed at Chaeronea, and the second, of eighty thousand men, met a similar fate at Orchomenus. At the same time, Lucullus, having collected all the maritime resources of the cities of Asia Minor, the islands, and particularly of Rhodes, was prepared to transport Sylla's army from Sestos to Asia; and Mithridates, from fear, made peace.

In the second and third wars, respectively conducted by Murena and Lucullus, there were no descents effected. Mithridates, driven step by step into Colchis, and no longer able to keep the sea, conceived the project of turning the Black Sea by the Caucasus, in order to pass through Thrace to assume the offensive,--a policy which it is difficult to understand, in view of the fact that he was unable to defend his kingdom against fifty thousand Romans.

Caesar, in his second descent on England, had six hundred vessels, transporting forty thousand men. During the civil wars he transported thirty-five thousand men to Greece. Antony came from Brundusium to join him with twenty thousand men, and passed through the fleet of Pompey,--in which act he was as much favored by the lucky star of Caesar as by the arrangements of his lieutenants.

Afterward Caesar carried an army of sixty thousand men to Africa; they did not, however, go in a body, but in successive detachments.

The greatest armament of the latter days of the Roman republic was that of Augustus, who transported eighty thousand men and twelve thousand horses into Greece to oppose Antony; for, besides the numerous transports required for such an army, there were two hundred and sixty vessels of war to protect them. Antony was superior in force on land, but trusted the empire of the world to a naval battle: he had one hundred and seventy war-vessels, in addition to sixty of Cleopatra's galleys, the whole manned by twenty-two thousand choice troops, besides the necessary rowers.

Later, Germanicus conducted an expedition of one thousand vessels, carrying sixty thousand men, from the mouths of the Rhine to the mouths of the Ems. Half of this fleet was destroyed on its return by a storm; and it is difficult to understand why Germanicus, controlling both banks of the Rhine, should have exposed his army to the chances of the sea, when he could have reached the same point by land in a few days.

When the Roman authority extended from the Rhine to the Euphrates, maritime expeditions were rare; and the great contest with the races of the North of Europe, which began after the division of the empire, gave employment to the Roman armies on the sides of Germany and Thrace. The eastern fraction of the empire still maintained a powerful navy, which the possession of the islands of the Archipelago made a necessity, while at the same time it afforded the means.

The first five centuries of the Christian era afford but few events of interest in maritime warfare. The Vandals, having acquired Spain, landed in Africa, eighty thousand strong, under Genseric. They were defeated by Belisarius; but, holding the Balearic Isles and Sicily, they controlled the Mediterranean for a time.

At the very epoch when the nations of the East invaded Europe, the Scandinavians began to land on the coast of England. Their operations are little better known than those of the barbarians: they are hidden in the mysteries of Odin.

The Scandinavian bards attribute two thousand five hundred vessels to Sweden. Less poetical accounts assign nine hundred and seventy to the Danes and three hundred to Norway: these frequently acted in concert.

The Swedes naturally turned their attention to the head of the Baltic, and drove the Varangians into Russia. The Danes, more favorably situated with respect to the North Sea, directed their course toward the coasts of France and England.

If the account cited by Depping is correct, the greater part of these vessels were nothing more than fishermen's boats manned by a score of rowers. There were also *snekars*, with twenty banks or forty rowers. The largest had thirty-four banks of rowers. The incursions of the Danes, who had long before ascended the Seine and Loire, lead us to infer that the greater part of these vessels were very small.

However, Hengist, invited by the Briton Vortigern, transported five thousand Saxons to England in eighteen vessels,--which would go to show that there were then also large vessels, or that the marine of the Elbe was superior to that of the Scandinavians.

Between the years 527 and 584, three new expeditions, under Ida and Cridda, gained England for the Saxons, who divided it into seven kingdoms; and it was not until three centuries had elapsed (833) that they were again united under the authority of Egbert.

The African races, in their turn, visited the South of Europe. In 712, the Moors crossed the Straits of Gibraltar, under the lead of Tarik. They came, five thousand strong, at the invitation of Count Julian; and, far from meeting great resistance, they were welcomed by the numerous enemies of the Visigoths. This was the happy era of the Caliphs, and the Arabs might well pass for liberators in comparison with the tyrants of the North. Tarik's army, soon swelled to twenty thousand men, defeated Rodrigo at Jerez and reduced the kingdom to submission. In time, several millions of the inhabitants of Mauritania crossed the sea and settled in Spain; and if their numerous migrations cannot be regarded as descents, still, they form one of the most curious and interesting scenes in history, occurring between the incursions of the Vandals in Africa and the Crusades in the East.

A revolution not less important, and one which has left more durable traces, marked in the North the establishment of the vast empire now known as Russia. The Varangian princes, invited by the Novgorodians, of whom Rurik was the chief, soon signalized themselves by great expeditions.

In 902, Oleg is said to have embarked eighty thousand men in two thousand boats on the Dnieper: they passed the falls of the river and debouched in the Black Sea, while their cavalry followed the banks. They proceeded to Constantinople, and forced Leo the Philosopher to pay tribute.

Forty years subsequently, Igor took the same route with a fleet said to have consisted of ten thousand boats. Near Constantinople his fleet, terrified by the effects of the Greek fire, was driven on the coast of Asia, where the force was disembarked. It was defeated, and the expedition returned home.

Not discouraged, Igor re-established his fleet and army and descended to the mouths of the Danube, where the Emperor Romanus I. sent to renew the tribute and ask for peace, (943.)

In 967, Svatoslav, favored by the quarrel of Nicephorus with the King of Bulgaria, embarked sixty thousand men, debouched into the Black Sea, ascended the Danube, and seized Bulgaria. Recalled by the Petchenegs, who were menacing Kiew, he entered into alliance with them and returned into Bulgaria, broke his alliance with the Greeks, and, being reinforced by the Hungarians, crossed the Balkan and marched to attack Adrianople. The throne of Constantine was held by Zimisces, who was worthy of his position. Instead of purchasing safety by paying tribute, as his predecessors had done, he raised one hundred thousand men, armed a respectable fleet, repulsed Svatoslav at Adrianople, obliged him to retreat to Silistria, and took by assault the capital of the Bulgarians. The Russian prince marched to meet him, and gave battle not far from Silistria, but was obliged to re-enter the place, where he sustained one of the most memorable sieges recorded in history.

In a second and still more bloody battle, the Russians performed prodigies of valor, but were again compelled to yield to numbers. Zimisces, honoring courage, finally concluded an advantageous treaty.

About this period the Danes were attracted to England by the hope of pillage; and we are told that Lothaire called their king, Ogier, to France to be avenged of his brothers. The first success of these pirates increased their fondness for this sort of adventure, and for five or six years their bands swarmed on the coasts of France and Britain and devastated the country. Ogier, Hastings, Regner, and Sigefroi conducted them sometimes to the mouths of the Seine, sometimes to the mouths of the Loire, and finally to those of the Garonne. It is even asserted that Hastings entered the Mediterranean and ascended the Rhone to Avignon; but this is, to say the least, doubtful. The strength of their fleets is not known: the largest seems to have been of three hundred sail.

In the beginning of the tenth century, Rollo at first landed in England, but, finding little chance of success against Alfred, he entered into alliance with him, landed in Neustria in 911, and advanced from Rouen on Paris: other bodies marched from Nantes on Chartres. Repulsed here, Rollo overran and ravaged the neighboring provinces. Charles the Simple saw no better means of delivering his kingdom of this ever-increasing scourge than to offer Rollo the fine province of Neustria on condition that he would marry his daughter and turn Christian,--an offer which was eagerly accepted.

Thirty years later, Rollo's step-son, annoyed by the successors of Charles, called to his aid the King of Denmark. The latter landed in considerable force, defeated the French, took the king prisoner, and assured Rollo's son in the possession of Normandy.

During the same interval (838 to 950) the Danes exhibited even greater hostility toward England than to France, although they were much more assimilated to the Saxons than to the French in language and customs. Ivar, after pillaging the kingdom, established his family in Northumberland. Alfred the Great, at first beaten by Ivar's successors, succeeded in regaining his throne and in compelling the submission of the Danes.

The aspect of affairs changes anew: Sweyn, still more fortunate than Ivar, after conquering and devastating England, granted peace on condition that a sum of money should be paid, and returned to Denmark, leaving a part of his army behind him.

Ethelred, who had weakly disputed with Sweyn what remained of the Saxon power, thought he could not do better to free himself from his importunate guests than to order a simultaneous massacre of all the Danes in the kingdom, (1002.) But Sweyn reappeared in the following year at the head of an imposing force, and between 1003 and 1007 three successive fleets effected disembarkations on the coast, and unfortunate England was ravaged anew.

In 1012, Sweyn landed at the mouth of the Humber and again swept over the land like a torrent, and the English, tired of obedience to kings who could not defend them, recognized him as king of the North. His son, Canute the Great, had to contend with a rival more worthy of him, (Edmund Ironside.) Returning from Denmark at the head of a considerable force, and aided by the perfidious Edric, Canute ravaged the southern part of England and threatened London. A new division of the kingdom resulted; but, Edmund having been assassinated by Edric, Canute was finally recognized as king of all England. Afterward he sailed to conquer Norway, from which country he returned to attack Scotland. When he died, he divided the kingdom between his three children, according to the usage of the times.

Five years after Canute's death, the English assigned the crown to their Anglo-Saxon princes; but Edward, to whom it fell, was better fitted to be a monk than to save a kingdom a prey to such commotions. He died in 1066, leaving to Harold a crown which the chief of the Normans settled in France contested with him, and to whom, it is said, Edward had made a cession of the kingdom. Unfortunately for Harold, this chief was a great and ambitious man.

The year 1066 was marked by two extraordinary expeditions. While William the Conqueror was preparing in Normandy a formidable armament against Harold, the brother of the latter, having been driven from Northumberland for his crimes, sought support in Norway, and, with the King of Norway, set out with thirty thousand men on five hundred vessels, and landed at the mouth of the Humber. Harold almost entirely destroyed this force in a bloody battle fought near York; but a more formidable storm was about to burst upon his head. William took advantage of the time when the Anglo-Saxon king was fighting the Norwegians, to sail from St. Valery with a very large armament. Hume asserts that he had three thousand transports; while other authorities reduce the number to twelve hundred, carrying from sixty to seventy thousand men. Harold hastened from York, and fought a decisive battle near Hastings, in which he met an honorable death, and his fortunate rival soon reduced the country to submission.

At the same time, another William, surnamed Bras-de-fer, Robert Guiscard, and his brother Roger, conquered Calabria and Sicily with a handful of troops,(1058 to 1070.)

Scarcely thirty years after these memorable events, an enthusiastic priest animated Europe with a fanatical frenzy and precipitated large forces upon Asia to conquer the Holy Land.

At first followed by one hundred thousand men, afterward by two hundred thousand badly-armed vagabonds who perished in great part under the attacks of the Hungarians, Bulgarians, and Greeks, Peter the Hermit succeeded in crossing the Bosporus, and arrived before Nice with from fifty to sixty thousand men, who were either killed or captured by the Saracens.

An expedition more military in its character succeeded this campaign of religious pilgrims. One hundred thousand men, composed of French, Burgundians, Germans, and inhabitants of Lorraine, under Godfrey of Bouillon, marched through Austria on Constantinople; an equal number, under the Count of Toulouse, marched by Lyons, Italy, Dalmatia, and Macedonia; and Bohemond, Prince of Tarentum, embarked with a force of Normans, Sicilians, and Italians, and took the route by Greece on Gallipolis.

This extensive migration reminds us of the fabulous expeditions of Xerxes. The Genoese, Venetian, and Greek fleets were chartered to transport these swarms of Crusaders by the Bosporus or Dardanelles to Asia. More than four hundred thousand men were concentrated on the plains of Nice, where they avenged the defeat of their predecessors. Godfrey afterward led them across Asia and Syria as far as Jerusalem, where he founded a kingdom.

All the maritime resources of Greece and the flourishing republics of Italy were required to transport these masses across the Bosporus and in provisioning them during the siege of Nice; and the great impulse thus given to the coast states of Italy was perhaps the most advantageous result of the Crusades.

This temporary success of the Crusaders became the source of great disasters. The Mussulmans, heretofore divided among themselves, united to resist the infidel, and divisions began to appear in the Christian camps. A new expedition was necessary to aid the kingdom which the brave Noureddin was threatening. Louis VII. and the Emperor Conrad, each at the head of one hundred thousand Crusaders, marched, as their predecessors had done, by the route of Constantinople, (1142.) But the Greeks, frightened by the recurring visits of these menacing guests, plotted their destruction.

Conrad, who was desirous of being first, fell into the traps laid for him by the Turks, and was defeated in detachments in several battles by the Sultan of Iconium. Louis, more fortunate, defeated the Turks on the banks of the Mender; but, being deprived of the support of Conrad, and his army being annoyed and partially beaten by the enemy in the passage of defiles, and being in want of supplies, he was confined to Attalia, on the coast of Pamphylia, where he endeavored to embark his army. The means furnished by the Greeks were insufficient, and not more than fifteen or twenty thousand men arrived at Antioch with the king: the remainder either perished or fell into the hands of the Saracens.

This feeble reinforcement soon melted away under the attacks of the climate and the daily contests with the enemy, although they were continually aided by small bodies brought over from Europe by the Italian ships; and they were again about to yield under the attacks of Saladin, when the court of Rome succeeded in effecting an alliance between the Emperor Frederick Barbarossa and the Kings of France and England to save the Holy Land.

The emperor was the first to set out. At the head of one hundred thousand Germans, he opened a passage through Thrace in spite of the formal resistance of the Greeks, now governed by Isaac Angelus. He marched to Gallipolis, crossed the Dardanelles, and seized Iconium. He died in consequence of an imprudent bath in a river, which, it has been pretended, was the Cydnus. His son, the Duke of Swabia, annoyed by the Mussulmans and attacked by diseases, brought to Ptolemais scarcely six thousand men.

At the same time, Richard Coeur-de-Lion[1] and Philip Augustus more judiciously took the route over the sea, and sailed from Marseilles and Genoa with two immense fleets,(1190.) The first seized Cyprus, and both landed in Syria,--where they would probably have triumphed but for the rivalry which sprang up between them, in consequence of which Philip returned to France.

Twelve years later, a new Crusade was determined upon, (1203.) Part of the Crusaders embarked from Provence or Italy; others, led by the Count of Flanders and the Marquis of Montferrat, proceeded to Venice, with the intention of embarking there. The party last mentioned were persuaded by the skillful Dandolo to aid him in an attack upon Constantinople, upon the pretext of upholding the rights of Alexis Angelus, the son of Isaac Angelus, who had fought the Emperor Frederick and was the successor of those Comnenuses who had connived at the destruction of the armies of Conrad and Louis VII.

---

[1] Richard sailed from England with twenty thousand foot and five thousand horsemen, and landed in Normandy, whence he proceeded by land to Marseilles. We do not know what fleet he employed to transport his troops to Asia. Philip embarked at Genoa on Italian ships, and with a force at least as large as that of Richard.

Twenty thousand men had the boldness to attack the ancient capital of the world, which had at least two hundred thousand defenders. They assailed it by sea and land, and captured it. The usurper fled, and Alexis was replaced upon the throne, but was unable to retain his seat: the Greeks made an insurrection in favor of Murzupha, but the Latins took possession of Constantinople after a more bloody assault than the first, and placed upon the throne their chief, Count Baldwin of Flanders. This empire lasted a half-century. The remnant of the Greeks took refuge at Nice and Trebizond.

A sixth expedition was directed against Egypt by John of Brienne, who, notwithstanding the successful issue of the horrible siege of Damietta, was obliged to give way before the constantly-increasing efforts of the Mussulman population. The remains of his splendid army, after a narrow escape from drowning in the Nile, deemed themselves very fortunate in being able to purchase permission to re-embark for Europe.

The court of Rome, whose interest it was to keep up the zeal of Christendom in these expeditions, of which it gathered all the fruits, encouraged the German princes to uphold the tottering realm at Jerusalem. The Emperor Frederick and the Landgrave of Hesse embarked at Brundusium in 1227, at the head of forty thousand chosen soldiers. The landgrave, and afterward Frederick himself, fell sick, and the fleet put in at Tarentum, from which port the emperor, irritated by the presumption of Gregory IX., who excommunicated him because he was too slow in the gratification of his wishes, at a later date proceeded with ten thousand men, thus giving way to the fear inspired by the pontifical thunders.

Louis IX., animated by the same feeling of fear, or impelled, if we may credit Ancelot, by motives of a higher character, set out from Aigues-Mortes, in 1248, with one hundred and twenty large vessels, and fifteen hundred smaller boats, hired from the Genoese, the Venetians and the Catalans; for France was at that time without a navy, although washed by two seas. This king proceeded to Cyprus, and, having there collected a still larger force, set out, according to Joinville's statement, with more than eighteen hundred vessels, to make a descent into Egypt. His army must have numbered about eighty thousand men; for, although half of the fleet was scattered and cast away upon the coast of Syria, he marched upon Cairo a few months later with sixty thousand fighting-men, twenty thousand being mounted. It should be stated that the Count of Poictiers had arrived also with troops from France.

The sad fortune experienced by this splendid army did not prevent the same king from engaging in a new Crusade, twenty years later,(1270.) He disembarked upon that occasion at the ruins of Carthage, and besieged Tunis. The plague swept off half his army in a few months, and himself was one of its victims. The King of Sicily, having arrived with powerful reinforcements at the time of Louis's death, and desiring to carry back the remains of the army to his island of Sicily, encountered a tempest which caused a loss of four thousand men and twenty large ships. This prince was not deterred by this misfortune from desiring the conquest of the Greek empire and of Constantinople, which seemed a prize of greater value and more readily obtained. Philip, the son and successor of Saint Louis, being anxious to return to France, would have nothing to do with that project. This was the last effort. The Christians who were abandoned in Syria were destroyed in the noted attacks of Tripoli and Ptolemais: some of the remnants of the religious orders took refuge at Cyprus and established themselves at Rhodes.

The Mussulmans, in their turn, crossed the Dardanelles at Gallipolis in 1355, and took possession, one after the other, of the European provinces of the Eastern Empire, to which the Latins had themselves given the fatal blow.

Mohammed II., while besieging Constantinople in 1453, is said to have had his fleet transported by land with a view to placing it in the canal and closing the port: it is stated to have been large enough to be manned by twenty thousand select foot-soldiers. After the capture of this capital, Mohammed found his means increased by all those of the Greek navy, and in a short time his empire attained the first rank of maritime powers. He ordered an attack to be made upon Rhodes and upon Otranto on the Italian main, whilst he proceeded to Hungary in search of a more worthy opponent (Hunniades.) Repulsed and wounded at Belgrade, the sultan fell upon Trebizond with a numerous fleet, brought that city to sue for terms, and then proceeded with a fleet of four hundred sail to make a landing upon the island of Negropont, which he carried by assault. A second attempt upon Rhodes, executed, it is stated, at the head of a hundred thousand men, by one of his ablest lieutenants, was a failure, with loss to the assailants. Mohammed was preparing to go to that point himself with an immense army assembled on the shores of Ionia, which Vertot estimates at three hundred thousand men; but death closed his career, and the project was not carried into effect.

About the same period England began to be formidable to her neighbors on land as well as on the sea; the Dutch also, reclaiming their country from the inroads of the sea, were laying the foundations of a power more extraordinary even than that of Venice.

Edward III. landed in France and besieged Calais with eight hundred ships and forty thousand men.

Henry V. made two descents in 1414 and 1417: he had, it is stated, fifteen hundred vessels and only thirty thousand men, of whom six thousand were cavalry.

All the events we have described as taking place, up to this period, and including the capture of Constantinople, were before the invention of gunpowder; for if Henry V. had cannon at Agincourt, as is claimed by some writers, they were certainly not used in naval warfare. From that time all the combinations of naval armaments were entirely changed; and this revolution took place--if I may use that expression--at the time when the invention of the mariner's compass and the discovery of America and of the Cape of

Good Hope were about to turn the maritime commerce of the world into new channels and to establish an entirely new system of colonial dependencies.

I shall not mention in detail the expeditions of the Spaniards to America, or those of the Portuguese, Dutch, and English to India by doubling the Cape of Good Hope. Notwithstanding their great influence upon the commerce of the world,--notwithstanding the genius of Gama, Albuquerque, and Cortez,--these expeditions, undertaken by small bodies of two or three thousand men against tribes who knew nothing of fire-arms, are of no interest in a military point of view.

The Spanish navy, whose fame had been greatly increased by this discovery of a new world, was at the height of its splendor in the reign of Charles V. However, the glory of the expedition to Tunis, which was conquered by this prince at the head of thirty thousand fine soldiers transported in five hundred Genoese or Spanish vessels, was balanced by the disaster which befell a similar expedition against Algiers, (1541,) undertaken when the season was too far advanced and in opposition to the wise counsels of Admiral Doria. The expedition was scarcely under way when the emperor saw one hundred and sixty of his ships and eight thousand men swallowed up by the waves: the remainder was saved by the skill of Doria, and assembled at Cape Metafuz, where Charles V. himself arrived, after encountering great difficulties and peril.

While these events were transpiring, the successors of Mohammed were not neglecting the advantages given them by the possession of so many fine maritime provinces, which taught them at once the importance of the control of the sea and furnished means for obtaining it. At this period the Turks were quite as well informed with reference to artillery and the military art in general as the Europeans. They reached the apex of their greatness under Solyman I., who besieged and captured Rhodes (1552) with an army stated to have reached the number of one hundred and forty thousand men,--which was still formidable even upon the supposition of its strength being exaggerated by one-half.

In 1565, Mustapha and the celebrated Dragut made a descent upon Malta, where the Knights of Rhodes had made a new establishment; they carried over thirty-two thousand Janissaries, with one hundred and forty ships. John of Valetta, as is well known, gained an enduring fame by repulsing them.

A more formidable expedition, consisting of two hundred vessels and fifty-five thousand men, was sent in 1527 to the isle of Cyprus, where Nicosia was taken and Famagosta besieged. The horrible cruelties practiced by Mustapha increased the alarm occasioned by his progress. Spain, Venice, Naples, and Malta united their naval forces to succor Cyprus; but Famagosta had already surrendered, notwithstanding the heroic defense of Bragadino, who was perfidiously flayed alive by Mustapha's order, to avenge the death of forty thousand Turks that had perished in the space of two years spent on the island.

The allied fleet, under the orders of two heroes, Don John of Austria, brother of Philip II., and Andrea Doria, attacked the Turkish fleet at the entrance of the Gulf of Lepanto, near the promontory of Actium, where Antony and Augustus once fought for the empire of the world. The Turkish fleet was almost entirely destroyed: more than two hundred vessels and thirty thousand Turks were captured or perished, (1571.) This victory did not put an end to the supremacy of the Turks, but was a great check in their career of greatness. However, they made such vigorous efforts that as large a fleet as the former one was sent to sea during the next year. Peace terminated this contest, in which such enormous losses were sustained.

The bad fortune of Charles V. in his expedition against Algiers did not deter Sebastian of Portugal from wishing to attempt the conquest of Morocco, where he was invited by a Moorish prince who had been deprived of his estates. Having disembarked upon the shores of Morocco at the head of twenty thousand men, this young prince was killed and his army cut to pieces at the battle of Alcazar by Muley Abdulmalek, in 1578.

Philip II., whose pride had increased since the naval battle of Lepanto on account of the success he had gained in France by his diplomacy and by the folly of the adherents of the League, deemed his arms irresistible. He thought to bring England to his feet. The invincible Armada intended to produce this effect, which has been so famous, was composed of an expeditionary force proceeding from Cadiz, including, according to Hume's narrative, one hundred and thirty-seven vessels, armed with two thousand six hundred and thirty bronze cannon, and carrying twenty thousand soldiers, in addition to eleven thousand sailors. To these forces was to be added an army of twenty-five thousand men which the Duke of Parma was to bring up from the Netherlands by way of Ostend. A tempest and the efforts of the English caused the failure of this expedition, which, although of considerable magnitude for the period when it appeared, was by no means entitled to the high-sounding name it received: it lost thirteen thousand men and half the vessels before it even came near the English coast.

After this expedition comes in chronological order that of Gustavus Adolphus to Germany,(1630.) The army contained only from fifteen to eighteen thousand men: the fleet was quite large, and was manned by nine thousand sailors; M. Ancillon must, however, be mistaken in stating that it carried eight thousand cannon. The debarkation in Pomerania received little opposition from the Imperial troops, and the King of Sweden had a strong party among the German people. His successor was the leader of a very extraordinary expedition, which is resembled by only one other example mentioned in history: I refer to the march of Charles X. of Sweden across the Belt upon the ice, with a view of moving from Sleswick upon Copenhagen by way of the island of Funen,(1658.) He had twenty-five thousand men, of whom nine thousand were cavalry, and artillery in proportion. This undertaking was so much the more rash because the ice was unsafe, several pieces of artillery and even the king's own carriage having broken through and been lost.

After seventy-five years of peace, the war between Venice and the Turks recommenced in 1645. The latter transported an army of fifty-five thousand men, in three hundred and fifty vessels, to Candia, and gained possession of the important post of Canea before the republic thought of sending succor. Although the people of Venice began to lose the spirit which made her great, she still numbered among her citizens some noble souls: Morosini, Grimani, and Mocenigo struggled several years against the Turks, who derived great advantages from their numerical superiority and the possession of Canea. The Venetian fleet had, nevertheless, gained a marked ascendency under the orders of Grimani, when a third of it was destroyed by a frightful tempest, in which the admiral himself perished.

In 1648, the siege of Candia began. Jussuf attacked the city furiously at the head of thirty thousand men: after being repulsed in two assaults, he was encouraged to attempt a third by a large breach being made. The Turks entered the place: Mocenigo rushed to meet them, expecting to die in their midst. A brilliant victory was the reward of his heroic conduct: the enemy were repulsed and the ditches filled with their dead bodies.

Venice might have driven off the Turks by sending twenty thousand men to Candia; but Europe rendered her but feeble support, and she had already called into active service all the men fit for war she could produce.

The siege, resumed some time after, lasted longer than that of Troy, and each campaign was marked by fresh attempts on the part of the Turks to carry succor to their army and by naval victories gained by the Venetians. The latter people had kept up with the advance of naval tactics in Europe, and thus were plainly superior to the Mussulmans, who adhered to the old customs, and were made to pay dearly for every attempt to issue from the Dardanelles. Three persons of the name of Morosini, and several Mocenigos, made themselves famous in this protracted struggle.

Finally, the celebrated Coprougli, placed by his merits at the head of the Ottoman ministry, resolved to take the personal direction of this war which had lasted so long: he accordingly proceeded to the island, where transports had landed fifty thousand men, at whose head he conducted the attack in a vigorous manner.(1667.)

In this memorable siege the Turks exhibited more skill than previously: their artillery, of very heavy caliber, was well served, and, for the first time, they made use of trenches, which were the invention of an Italian engineer.

The Venetians, on their side, greatly improved the methods of defense by mines. Never had there been seen such furious zeal exhibited in mutual destruction by combats, mines, and assaults. Their heroic resistance enabled the garrison to hold out during winter: in the spring, Venice sent reinforcements and the Duke of Feuillade brought a few hundreds of French volunteers.

The Turks had also received strong reinforcements, and redoubled their efforts. The siege was drawing to a close, when six thousand Frenchmen came to the assistance of the garrison under the leadership of the Duke of Beaufort and Navailles,(1669.) A badly-conducted sortie discouraged these presumptuous young men, and Navailles, disgusted with the sufferings endured in the siege, assumed the responsibility, at the end of two months, of carrying the remnant of his troops back to France. Morosini, having then but three thousand exhausted men to defend a place which was open on all sides, finally consented to evacuate it, and a truce was agreed upon, which led to a formal treaty of peace. Candia had cost the Turks twenty-five years of efforts and more than one hundred thousand men killed in eighteen assaults and several hundred sorties. It is estimated that thirty-five thousand Christians of different nations perished in the glorious defense of the place.

The struggle between Louis XIV., Holland, and England gives examples of great maritime operations, but no remarkable descents. That of James II. in Ireland (1690) was composed of only six thousand Frenchmen, although De Tourville's fleet contained seventy-three ships of the line, carrying five thousand eight hundred cannon and twenty-nine thousand sailors. A grave fault was committed in not throwing at least twenty thousand men into Ireland with such means as were disposable. Two years later, De Tourville had been conquered in the famous day of La Hogue, and the remains of the troops which had landed were enabled to return through the instrumentality of a treaty which required their evacuation of the island.

At the beginning of the eighteenth century, the Swedes and Russians undertook two expeditions very different in character.

Charles XII., wishing to aid the Duke of Holstein, made a descent upon Denmark at the head of twenty thousand men, transported by two hundred vessels and protected by a strong squadron. He was really assisted by the English and Dutch navies, but the expedition was not for that reason the less remarkable in the details of the disembarkation. The same prince effected a descent into Livonia to aid Narva, but he landed his troops at a Swedish port.

Peter the Great, having some cause of complaint against the Persians, and wishing to take advantage of their dissensions, embarked (in 1722) upon the Volga: he entered the Caspian Sea with two hundred and seventy vessels, carrying twenty thousand foot-soldiers, and descended to Agrakhan, at the mouths of the Koisou, where he expected to meet his cavalry. This force, numbering nine thousand dragoons and five thousand Cossacks, joined him after a land-march by way of the Caucasus. The czar then seized Derbent, besieged Bakou, and finally made a treaty with one of the parties whose dissensions at that time filled with discord the empire of the Soofees: he procured the cession of Astrabad, the key of the Caspian Sea and, in some measure, of the whole Persian empire.

The time of Louis XV. furnished examples of none but secondary expeditions, unless we except that of Richelieu against Minorca, which was very glorious as an escalade, but less extraordinary as a descent.

[In 1762, an English fleet sailed from Portsmouth: this was joined by a portion of the squadron from Martinico. The whole amounted to nineteen ships of the line, eighteen smaller vessels of war, and one hundred and fifty transports, carrying ten thousand men. The expedition besieged and captured Havana.--TRS.]

The Spaniards, however, in 1775, made a descent with fifteen or sixteen thousand men upon Algiers, with a view of punishing those rovers of the sea for their bold piracies; but the expedition, for want of harmonious action between the squadron and the land-forces, was unsuccessful, on account of the murderous fire which the troops received from the Turkish and Arab musketeers dispersed among the undergrowth surrounding the city. The troops returned to their vessels after having two thousand men placed *hors de combat.*

The American war (1779) was the epoch of the greatest maritime efforts upon the part of the French. Europe was astonished to see this power send Count d'Estaing to America with twenty-five ships of the line, while at the same time M. Orvilliers, with a Franco-Spanish fleet of sixty-five ships of the line, was to cover a descent to be effected with three hundred transports and forty thousand men, assembled at Havre and St. Malo.

This new armada moved back and forth for several months, but accomplished nothing: the winds finally drove it back to port.

D'Estaing was more fortunate, as he succeeded in getting the superiority in the Antilles and in landing in the United States six thousand Frenchmen under Rochambeau, who were followed, at a later date, by another division, and assisted in investing the English army under Cornwallis at Yorktown, (1781:) the independence of America was thus secured. France would perhaps have gained a triumph over her implacable rival more lasting in its effects, had she, in addition to the display made in the English Channel, sent ten ships and seven or eight thousand men more to India with Admiral Suffren.

During the French Revolution, there were few examples of descents: the fire at Toulon, emigration, and the battle of Ushant had greatly injured the French navy.

Hoche's expedition against Ireland with twenty-five thousand men was scattered by the winds, and no further attempts in that quarter were made. (1796.)

At a later date, Bonaparte's expedition to Egypt, consisting of twenty-three thousand men, thirteen ships, seventeen frigates, and four hundred transports, obtained great successes at first, which were followed by sad reverses. The Turks, in hopes of expelling him, landed fifteen thousand men at Aboukir, but were all captured or driven into the sea, notwithstanding the advantages this peninsula gave them of intrenching themselves and waiting for reinforcements. This is an excellent example for imitation by the party on the defensive under similar circumstances.

The expedition of considerable magnitude which was sent out in 1802 to St. Domingo was remarkable as a descent, but failed on account of the ravages of yellow fever.

Since their success against Louis XIV., the English have given their attention more to the destruction of rival fleets and the subjugation of colonies than to great descents. The attempts made in the eighteenth century against Brest and Cherbourg with bodies of ten or twelve thousand men amounted to nothing in the heart of a powerful state like France. The remarkable conquests which procured them their Indian empire occurred in succession. Having obtained possession of Calcutta, and then of Bengal, they strengthened themselves gradually by the arrival of troops in small bodies and by using the Sepoys, whom they disciplined to the number of one hundred and fifty thousand.

The Anglo-Russian expedition to Holland in 1799 was composed of forty thousand men, but they were not all landed at once: the study of the details of the operations is, however, quite interesting.

In 1801, Abercrombie, after threatening Ferrol and Cadiz, effected a descent into Egypt with twenty thousand Englishmen. The results of this expedition are well known.

General Stuart's expedition to Calabria, (1806,) after some successes at Maida, was for the purpose of regaining possession of Sicily. That against Buenos Ayres was more unfortunate in its results, and was terminated by a capitulation.

In 1807, Lord Cathcart attacked Copenhagen with twenty-five thousand men, besieged and bombarded the city, and gained possession of the Danish fleet, which was his object.

In 1808, Wellington appeared in Portugal with fifteen thousand men. After gaining the victory of Vimeira, and assisted by the general rising of the Portuguese, he forced Junot to evacuate the kingdom. The same army, increased in numbers to twenty-five thousand and placed under Moore's command, while making an effort to penetrate into Spain with a view of relieving Madrid, was forced to retreat to Corunna and there re-embark, after suffering severe losses. Wellington, having effected another landing in Portugal with reinforcements, collected an army of thirty thousand Englishmen and as many Portuguese, with which he avenged Moore's misfortunes by surprising Soult at Oporto, (May, 1809,) and then beating Joseph at Talavera, under the very gates of his capital.

The expedition to Antwerp in the same year was one of the largest England has undertaken since the time of Henry V. It was composed of not less than seventy thousand men in all,--forty thousand land-forces and thirty thousand sailors. It did not succeed, on account of the incapacity of the leader.

A descent entirely similar in character to that of Charles X. of Sweden was effected by thirty Russian battalions passing the Gulf of Bothnia on the ice in five columns, with their artillery. Their object was to take possession of the islands of Aland and spread a feeling of apprehension to the very gates of Stockholm. Another division passed the gulf to Umea, (March, 1809.)

General Murray succeeded in effecting a well-planned descent in the neighborhood of Tarragona in 1813, with the intention of cutting Suchet off from Valencia: however, after some successful operations, he thought best to re-embark.

The expedition set on foot by England against Napoleon after his return from Elba in 1815 was remarkable on account of the great mass of *materiel* landed at Ostend and Antwerp. The Anglo-Hanoverian army contained sixty thousand men, but some came by land and others were disembarked at a friendly port.

The English engaged in an undertaking in the same year which may be regarded as very extraordinary: I refer to the attack on the capital of the United States. The world was astonished to see a handful of seven or eight thousand Englishmen making their appearance in the midst of a state embracing ten millions of people, taking possession of its capital, and destroying all the public buildings,--results unparalleled in history. We would be tempted to despise the republican and unmilitary spirit of the inhabitants of those states if the same militia had not risen, like those of Greece, Rome, and Switzerland, to defend their homes against still more powerful attacks, and if, in the same year, an English expedition more extensive than the other had not been entirely defeated by the militia of Louisiana and other states under the orders of General Jackson.

If the somewhat fabulous numbers engaged in the irruption of Xerxes and the Crusades be excepted, no undertaking of this kind which has been actually carried out, especially since fleets have been armed with powerful artillery, can at all be compared with the gigantic project and proportionate preparations made by Napoleon for throwing one hundred and fifty thousand veterans upon the shores of England by the use of three thousand launches or large gun-boats, protected by sixty ships of the line.[1]

From the preceding narrative the reader will perceive what a difference there is in point of difficulty and probability of success between descents attempted across a narrow arm of the sea, a few miles only in width, and those in which the troops and *materiel* are to be transported long distances over the open sea. This fact gives the reason why so many operations of this kind have been executed by way of the Bosporus.

---

[1]   See the account of the expedition to the Crimea.--TRANSLATORS.

The following paragraphs have been compiled from authentic data:–

In 1830, the French government sent an expedition to Algiers, composed of an army of thirty-seven thousand five hundred men and one hundred and eighty pieces of artillery. More than five hundred vessels of war and transports were employed. The fleet sailed from Toulon.

In 1838, France sent a fleet of twenty-two vessels to Vera Cruz. The castle of San Juan d'Ulloa fell into their hands after a short bombardment. A small force of about one thousand men, in three columns, took the city of Vera Cruz by assault: the resistance was slight.

In 1847, the United States caused a descent to be made upon the coast of Mexico, at Vera Cruz, with an army of thirteen thousand men, under the command of General Scott. One hundred and fifty vessels were employed, including men-of-war and transports. The city of Vera Cruz and the castle of San Juan d'Ulloa speedily fell into the possession of the forces of the United States. This important post became the secondary base of operations for the brilliant campaign which terminated with the capture of the city of Mexico.

In 1854 commenced the memorable and gigantic contest between Russia on the one side and England, France, Sardinia, and Turkey on the other. Several descents were made by the allied forces at different points of the Russian coast: of these the first was in the Baltic Sea. An English fleet sailed from Spithead, under the command of Sir Charles Napier, on the 12th of March, and a French fleet from Brest, under the command of Vice-Admiral Parseval Deschenes, on the 19th of April. They effected a junction in the Bay of Barosund on the 11th of June. The allied fleet numbered thirty ships and fifty frigates, corvettes, and other vessels. The naval commanders wished to attack the defenses of Bomarsund, on one of the Aland Isles, but, after a reconnoissance, they came to the conclusion that it was necessary to have land-forces. A French corps of ten thousand men was at once dispatched to Bomarsund under General Baraguay-d'Hilliers, and the place was speedily reduced.

Later in the same year, the great expedition to the Crimea was executed; and with reference to it the following facts are mentioned, in order to give an idea of its magnitude:–

September 14, 1854, an army of fifty-eight thousand five hundred men and two hundred pieces of artillery was landed near Eupatoria, composed of thirty thousand French, twenty-one thousand five hundred English, and seven thousand Turks. They were transported from Varna to the place of landing by three hundred and eighty-nine ships, steamers, and transports. This force fought and gained the battle of the Alma, (September 20,) and thence proceeded to Sebastopol. The English took possession of the harbor of Balaklava and the French of Kamiesch: these were the points to which subsequent reinforcements and supplies for the army in the Crimea were sent.

November 5, at the battle of Inkermann, the allied army numbered seventy-one thousand men.

At the end of January, 1855, the French force was seventy-five thousand men and ten thousand horses. Up to the same time, the English had sent fifty-four thousand men to the Crimea, but only fifteen thousand were alive, present, and fit for duty.

February 4, the French numbered eighty-five thousand; the English, twenty-five thousand fit for duty; the Turks, twenty-five thousand.

May 8, 1855, General La Marmora arrived at Balaklava with fifteen thousand Sardinians.

In the latter part of May, an expedition of sixteen thousand men was sent to Kertch.

In August, the French force at Sebastopol had risen to one hundred and twenty thousand men.

September 8, the final assault took place, which resulted in the evacuation of the place by the Russians. The allies had then in battery more than eight hundred pieces of artillery.

The fleet which co-operated with the land-forces in the artillery attack of October 17, 1854, consisted of twenty-five ships. There were present and prepared to attack in September, 1855, thirty-four ships.

October, 1855, an expeditionary force of nine thousand men was sent to Kinburn, which place was captured.

Marshal Vaillant, in his report, as Minister of War, to the French emperor, says there were sent from France and Algeria three hundred and ten thousand men and forty thousand horses, of which two hundred and twenty-seven thousand men returned to France and Algeria.

The marshal's report gives the following striking facts, (he refers only to French operations:-)

The artillery *materiel* at the disposal of the Army of the East comprised one thousand seven hundred guns, two thousand gun-carriages, two thousand seven hundred wagons, two millions of projectiles, and nine million pounds of powder. There were sent to the army three thousand tons of powder, seventy millions of infantry-cartridges, two hundred and seventy thousand rounds of fixed ammunition, and eight thousand war-rockets.

On the day of the final assault there were one hundred and eighteen batteries, which during the siege had consumed seven million pounds of powder. They required one million sand-bags and fifty thousand gabions.

Of engineer materials, fourteen thousand tons were sent. The engineers executed fifty miles of trenches, using eighty thousand gabions, sixty thousand fascines, and one million sand-bags.

Of subsistence, fuel, and forage, five hundred thousand tons were sent.

Of clothing, camp-equipage, and harness, twelve thousand tons.

Hospital stores, six thousand five hundred tons.

Provision-wagons, ambulances, carts, forges, &c, eight thousand tons.

In all, about six hundred thousand tons.

It is not thought necessary to add similar facts for the English, Sardinian, and Turkish armies.

In 1859, the Spaniards made a descent upon Morocco with a force of forty thousand infantry, eleven squadrons of cavalry, and eighty pieces of artillery, using twenty-one vessels of war with three hundred and twenty-seven guns, besides twenty-four gun-boats and numerous transports.

In 1860, a force of English and French was landed on the coast of China, whence they marched to Pekin and dictated terms of peace. This expedition is remarkable for the smallness of the numbers which ventured, at such a great distance from their sources of supply and succor, to land upon a hostile shore and penetrate into the midst of the most populous empire in the world.

The French expedition to Syria in 1860 was small in numbers, and presented no remarkable features.

Toward the close of the year 1861, the government of the United States sent an expedition of thirteen thousand men to Port Royal, on the coast of South Carolina, one of the seceding States. The fleet of war-vessels and transports sailed from Hampton Roads, under command of Captain Dupont, and was dispersed by a violent gale: the losses of men and *materiel* were small, however, and the fleet finally reached the rendezvous. The defenses of the harbor having been silenced by the naval forces, the disembarkation of the land-troops took place, General Sherman being in command.

England, France, and Spain are now (January 16, 1862) engaged in an expedition directed against Mexico. The first operations were the capture, by the Spanish forces, of Vera Cruz and its defenses: the Mexicans offered no resistance at that point. The future will develop the plans of the allies; but the ultimate result of a struggle (if, indeed, one be attempted by the Mexicans) cannot be doubted, when three of the most powerful states of Europe are arrayed against the feeble and tottering republic of Mexico.]

❀

CONCLUDING COMMENTARY

*J*omini's fall from favor was rapid, caused by the Prussian victory over the French in the Franco-Prussian War of 1870, a few years after his death. As soon as the Prussian system had proven itself in battle, the military world began looking to Germany as the epitome of martial excellence. Germany was in, and France was out; Clausewitz was in, and Jomini was out. Jomini's reputation fell victim to a change in fashion.

www.ingramcontent.com/pod-product-compliance
Lightning Source LLC
Chambersburg PA
CBHW030512100426
42813CB00001B/15